THE EVOLUTION MYTH

by Robert James Galgey

ISBN: 1451576080
ISBN-13: 9781451576085

CHAPTER 1

When Charles Darwin published his ORIGIN OF THE SPECIES in 1859, little did he know the controversy that would rage in the ensuing decades.

In the vernacular he opened up 'a can of worms.'

Contrary to popular opinion, Darwin only wrote his book as theory and had serious doubts about a number of aspects. To quote a writer on the LONDON TIMES, Christopher booker: "It was a beautifully simple and attractive theory. The only trouble was, that as Darwin was himself partly aware, it was full of colossal holes."

Surprisingly, many biologists cannot agree on the basics of evolution.

On this point note this comment in the scientific journal DISCOVER : "Evolution......is not only under attack by fundamentalist Christians, but it is also being questioned by reputable scientists. Among Paleontologists, scientists who study the fossil record, there is growing dissent from the prevailing view of Darwinism." (1)

Many scientists give lip service to the evolution theory because their career advancement depends on it. The people who decide their future expect it. Even though many of them harbour serious doubts, and some could be termed 'closet creation believers' they continue to profess belief in evolution due to peer pressure.

Because noted scientists claim to be believers, the average man in the street assumes it MUST be right. When famous documentary makers casually drop such expressions as : "millions of years ago this animal evolved....." in the midst of the most amazing photography, our minds can be easily swayed.

But have you ever noticed the voice of the narrator NEVER stops to explain HOW he arrived at this conclusion. The statement is made and the viewer just assumes it MUST be right coming from such an eloquent presenter. Surely he must have done his homework. Alas this is not the case. The only homework he's done is how to get the best angle, the best film of such an animal, bird or fish. And in this they deserve all credit as we can only wonder at how they manage to get the shots they do. But when it comes to explaining HOW such a creature evolved BASED ON THE FACTS this voice becomes mute.

The purpose of this book is to prove to the reader that the evolution theory is a myth without a single fact in support.

According to the dictionary, one definition of the word 'myth' is : "A widely held but false belief or fictitious thing." By a step by step analysis of the FACTS, this writer sets out to prove that the evolution theory, in whatever form it takes, is worthy of the above definition.

After accepting it as fact for most of your life, having been brainwashed into belief, you may be surprised at the avalanche of facts that are in direct conflict with the theory.

This has actually led to evolutionists creating a mythology all of their own. This was admitted by famous evolutionist Loren Eisley when he said :

"After having chided the theologian for his reliance on myth and miracle, science found itself in the unenviable position of having to create a mythology of its own : namely, the assumption that what, after long effort could not be proved to take place today, had in truth taken place in the primeval past." (2)

If this is true, then millions of people have been taken in by a lie. Our children are being taught this in school as though it were a fact. Some schools actually refuse to allow anything to be taught that might go counter to the theory. Not seeing any facts that might conflict with evolution prevents students from making an informed choice. Needless to say this is a very unhealthy situation where children are only taught one idea which is built on a tenuous foundation, to say the least.

I am not for a moment suggesting that evolution should be withdrawn from the curriculum, but that students should at least have the opportunity to have access to the scientific facts that might disagree with the theory.

There are so many questions that require answers, such as the following :

In view of discoveries made by Louis Pasteur in the 19th century is it possible for life to originate from inanimate matter ?

Is it feasible that the 20 amino acids necessary for life could form at random ?

Was the original atmosphere on earth friendly to the forming of a living cell ?

Seeing as proteins cannot be assembled without the nucleic acids (DNA & RNA) which cannot form without proteins, what came first ?

To find support, the evolution theory must be able to find a simple beginning. Has it done so ?

Because the theory requires the bridging of totally different species, have any links been found so far ?

Out of the millions of fossils to be found in museums and universities around the world, are any helpful to evolution?

If humanoids (part ape/ part man) really existed, how come they never left a trace of their existence, whereas apes are very much alive and well ?

Does the allowance of millions of years for evolution to take place fit the facts ? Is man as old as they say ?

Is the intelligence and physiology of apes really similar to man ?

We need the answers to these questions to be able to make an informed decision to either accept it or reject it. So where do we begin ?

Lets take a look at the evolutionists view of life originating from inanimate matter.

CHAPTER 2

Evolutionists make the assertion that the first speck of life on earth arose by itself from lifeless matter. Is this scientific ? For the answer we need to understand that all matter on earth is composed of basic elements.

What is an element ? It is a substance made up of atoms of one single kind and cannot be simplified or decomposed by ordinary chemical means. In other words its the very basic unit that makes up all matter on this planet.

This leads to the question : Is there any trend toward evolution of the elements ? The answer is no because atoms generally are stable or in some cases in a decaying state until they turn into an element that is stable.

This fact harmonizes with the scientific principle called ENTROPY, which means there is a tendency from the highly organized DOWNWARD toward the less organized.

We could illustrate this with the creation of a motor car. Would the elements that make up the vehicle, such as iron, aluminium, copper, silicon, tungsten etc., ever, left to themselves produce a motor car ? No, they would remain the way they are unless taken by man and fashioned into a machine such as a car. But even then, if left to itself it would eventually decay and return to the earth. Thus an example of entropy at work, whereby the tendency is from the highly organized DOWNWARD to the less organized. In actual fact entropy is the exact opposite of evolution.

There is a fundamental truth regarding inanimate matter on earth : It never searches out a way to improve itself but tends toward a state of neutralization or stability.

Any appeal to immense periods of time won't help either. Time is the enemy of evolution, it produces decay and disintegration. It corrodes metals, erodes cliffs, reduces the size of mountains. Time is DEstructive not CONstructive.

For many decades, scientists clung to the idea of spontaneous generation. Then along came Louis Pasteur the famous French scientist, who proved by experiments that life only comes from existing life, and that spontaneous generation is impossible.

Mathematical laws also gang up on the theory. As we will learn later, amino acids (the basic building blocks of life) exist in two forms: Left handed and right-handed. The twenty needed for life are all of the left-handed variety.

Its been estimated that the chances of all 20 coming together at random is 1 in 10 followed by 14,880 zero's. But this is where the problem becomes acute. These amino acids cannot be tied together indiscriminately, they must be in the right sequence. Mathematicians have calculated the odds of this happening as 1 in 10 followed by 79,360 zero's.

Dr. Emil Borel, an authority on probabilities, says that if there is less of a 1 in 10 followed by 50 zero's chance for something to happen, it will NEVER happen no matter how much time is allowed.

Yet all this evidence doesn't deter evolutionists from grasping at straws to keep their theory afloat. For example, knowing the extent of the problems facing the idea of spontaneous generation, some prominent evolutionists have tried to push them into outer space.

British astronomer Sir Fred Hoyle said that "existing terrestrial theories of the origin of life are highly unsatisfactory for sound chemical reasons" and that "life did not originate on earth itself, but rather on comets." But this idea doesn't explain how life originated without outside help, it only transfers the dilemma to another area.

Others believe it in spite of the mountain of evidence. Take for example Dr. George Wald, a Nobel prize-winning biologist. He admitted : "One only has to contemplate the magnitude of this task to concede that the spontaneous generation of a living organism is impossible. Yet here we are as a result I believe of spontaneous generation." (3)

Now just stop and think about what this man has said. He knows its impossible, but still believes it anyway. Surely this should ring alarm bells for those not part of the scientific community. A lot is at stake here. If the majority of the population continue to listen in awe and hang off every word of noted scientists, even when they present not the slightest shred of proof for their theory, and even admit its impossible, are they not subscribing to a lie? Do we really want to go through life blindly believing in what is a fable, a myth? Do we want our children to be taught false knowledge, which could in future prepare their acceptance of dishonesty on the printed page?

It is indeed an unhealthy situation, when an idea or proposition is promulgated with eagerness when it is known to be false. Its a sad day for the advance of true scientific knowledge.

So, where do we go from here? Lets just for a moment suspend any doubts and examine the evolutionists view of how life originated.

CHAPTER 3

Evolutionists speculate that millions of years ago, the earth had an atmosphere composed of carbon dioxide, methane, ammonia and water. They believe that energy coming from either the sun or lightening, broke up these compounds and then formed into amino acids which later united into what might be called protein. Eventually they speculate, all the seas must have become a sort of "organic soup" although lifeless.

From this soup, according to Richard Dawkins in his book THE SELFISH GENE : "A particularly remarkable molecule was formed by accident."   (4)

He believed that this molecule had the ability to reproduce itself. To try and prove that this was how it happened in the beginning, scientists have conducted experiments, creating an artificial "atmosphere" of hydrogen, methane, ammonia, and water vapour.

In 1953 a man by the name of Stanley Miller passed an electric spark through such an atmosphere and produced four of the necessary amino acids needed for life. He was 16 short of the 20 needed to build a protein.

In the years that followed, many scientists tried similar experiments but none have produced all 20 amino acids essential for life. Another problem they faced as did Miller was that the spark used to break up the chemicals in the atmosphere could quickly destroy the amino acids. To get around this, Miller rigged his experiment by using a trap in his apparatus which stored the acids as soon as they were formed.

Some scientists claim that the amino acids could have escaped the harmful effects of ultraviolet rays or lightening by plunging under the ocean. At this point another problem arises.

Water has a tendency to de-stabalize amino acids, making it hard for them to form in any quantity. Evolutionist Hitching explained this way: "Beneath the surface of the water there would not be enough energy to activate further chemical reactions. Water in any case inhibits growth of more complex molecules."

Hitching reveals another dilemma facing evolutionists: "With oxygen in the air the first amino acids would never have got started, without oxygen, it would have been wiped out by cosmic rays."   (5)

Let us for a moment sidestep these facts and imagine that somehow there was a 'primordial soup' friendly to millions of amino acids. What are the chances now of a protein forming in this environment ? Again the evolution theory is faced with a stubborn problem. I use the word stubborn but perhaps the word insurmountable would be more appropriate.

As mentioned earlier, of the 100 plus amino acids that exist only 20 are needed to make up life. Guess what ? They are all LEFT-HANDED. It remains a complete mystery to scientists why this is so, as there is no reason for it. But, be that as it may, it greatly increases the odds against all 20 forming at random. Added to this some proteins serve as enzymes which speed up chemical reactions in the living cell. On average 2000 perform this job. What are the chances of all 2000 uniting at random ? One chance in $10/40,000$ or 1 followed by 40,000 zero's. To put it in the parlance of a gambler, at the roll of the dice you would need to get 50,000 sixes in a row ! Now we all know this is impossible. Even Hoyle admitted: "An outrageously small probability, that could not be faced if the whole universe consisted of organic soup. "   (6)

The difficulties don't end there. Proteins cannot be assembled without the nucleic acids (DNA & RNA) nor can nucleic acids form without proteins. Its that old chicken and egg riddle, which came first ?   One scientist had this to say : "The origin of the genetic code poses a massive chicken and egg problem, that remains at present, completely scrambled." (7)

All along the scientists have dreamt of discovering a living cell that is super-simple; completely basic. Such a discovery would go a long way in support of evolution. Have they found such a cell ? A specialist in molecular biology, Michael Denton gives the answer: "Molecular biology has shown that even the simplest of all living systems on earth today, bacterial cells, are exceedingly complex objects. Although the tiniest bacterial cells are incredibly small, weighing less than $10/\_12$ grams, each is in effect a veritable micro-miniaturized factory containing thousands of exquisitely designed pieces of intricate molecular machinery, made up altogether of one hundred thousand million atoms, far more complicated than any machine built by man and absolutely without parallel in the non-living world."   (8)

Alas another dream bites the dust, because the living cell is a very complex thing. So much so, that the information in the DNA of a SINGLE cell "If written out would fill a thousand 600 page books."

Summarizing this chapter I would like to quote Fred Hoyle and his fellow scientist Chandra Wickramasinghe. They said: "The problem for biology is to reach a simple beginning......... fossil residues of ancient life forms discovered in the rocks do not reveal a simple beginning......so the evolution theory lacks a proper foundation".   (9)

If a large building could not stand without a proper foundation, how could the evolution theory be any different ? But for the moment let us suspend belief in the aforementioned facts and go to the next step proposed by evolutionists.

CHAPTER 4

The generally held view of the process of evolution is that life started as a single celled organism then into a fish, which in turn evolved into an amphibian, next a reptile, then a bird, which evolved into a mammal, then finally into a human.

This would all be very well if links could be readily found between each of the above. Alas this is not the case. This is why we frequently hear evolutionists speak of a "missing link".

For the theory to be true there must be evidence SOMEWHERE of transitional forms that bridge the gap between species. For example, if the fossil record showed a creature that was definitely half fish and half frog, this would give considerable support to evolution. Is the fossil record helpful in this regard ?

Francis Hitching supplies the answer: "Fish jump into the fossil record seemingly from nowhere : mysteriously, suddenly, full formed". (10)

For a fish to turn into an amphibian would require major modifications. Take for instance the above mentioned fish and frog. Their backbones are so different they would need to be changed beyond all recognition. The bones of their skulls are different. Fish fins would need to turn into jointed limbs having wrists and toes. Gills would have to become lungs. The fish has a two-chambered heart, an amphibian one with three chambers. Fish absorb sound through their bodies, whereas amphibians are endowed with eardrums. Amphibians have an extendable tongue, but no fish has such a tongue.

As you can see there are profound differences between these two types of creatures. But in spite of the foregoing facts, some evolutionists eagerly point to the lungfish as proof. Its true that the lungfish can actually breathe out of water for short periods, due to what is called a "swim bladder". So. Could this be the right candidate for linking fish and amphibians ? Not according to David Attenborough who disqualifies the fish : "Because the bones of their skulls are so different from those of the first fossil amphibians that the one cannot be derived from the other". (11)

Lets just suspend the foregoing facts and say, for the sake of argument that the fish somehow managed to evolve into an amphibian. What about the next step, from amphibian into reptile?

Unfortunately for evolutionists , the gulf between these two animals is very great indeed. For example amphibians lay soft jelly-like eggs in water which are fertilized EXTERNALLY. Although reptiles choose a watery environment to lay their eggs, they are still laid on land. The embryo of a reptile is nestled in a fluid environment but with a shell around it for protection. This means that the egg has to be fertilized INTERNALLY before the shell is formed. This calls for different sexual organs and mating methods which only illuminate the gulf between the amphibian and the reptile.

Another major difference is the disposal of waste. Amphibian embryos release their wastes in the surrounding water. If the urea in a reptile egg worked the same way it would kill the embryo. So how is this avoided ? A major chemical change takes place. The wastes which are uric acid are insoluble and are stored within an area called the allantois membrane. Its function is described in a book called THE REPTILES. It says : "The allantois receives and stores embryonic waste serving as a sort of bladder. It also has blood vessels that pick up oxygen that passes through the shell and conduct it to the embryo". (12)

These complex differences between these two creatures only highlight the fact they are not related in any way. The fossil record is not helpful either as evolutionist Archie Carr lamented: "One of the frustrating features of the fossil record of vertebrate history is that it shows so little about the evolution of reptiles during their earliest days, when the shelled egg was developing". (13)

For the moment lets again side-step the facts and go to the next step: The reptile evolving into a bird.

Most of us are aware of one outstanding characteristic of reptiles........they are cold-blooded creatures. This means that their internal temperature will either increase or decrease depending on the outside temperature. Birds in contrast are warm-blooded and maintain a relatively constant internal temperature regardless of what the weather is outside.

The belief that warm-blooded birds come from cold-blooded reptiles is inexplicable. It remains a mystery even to the most knowledgeable scientists. For example the French evolutionist Le Comte du Nouy said : "This stands out today as one of the great puzzles of evolution". (14)

Lets face it, how could such a difference be bridged without outside help ?

Evolutionists imagine that somehow the scales of a reptile evolved into the feathers of a bird. Is this scientific ? Well consider how complex the feather is in comparison to the scale and its uniqueness in the animal kingdom.

Feathers are amazing in design. Emanating from the shaft of a feather are what are called barbs. Each barb is connected by barbules together resembling a fishing net. Each of these barbules has hundreds of little barbicels and hooklets. The latter holds all the parts of the feather together to make vanes or flat surfaces. It has no equal as an airfoil.

A microscopic examination of a pigeon feather revealed it had several hundred thousand barbules and millions of barbicels and hooklets. A bird as big as a swan can have as many as 25,000 feathers.

We have often seen birds preening themselves with their beak and there is good reason for this. Preening helps for efficient flight. How do they do it ? They apply pressure on the feather and as the barbs pass through it causes the hooks and barbules to link together a bit like the teeth on a zipper.

In view of all this, does it not show the gulf between the scale of a reptile and a bird's feather? How could one possibly evolve into the other, it makes no sense.

Another difference in a bird over a reptile is the density of their bones. Naturally, to enable it to fly, the bird would require bones as light as possible while still strong, therefore their bones are thin and hollow; totally different to the bones of a reptile which are quite solid. What gives the bird's bones its strength are small struts inside the hollow bones which are a bit like the struts inside an aircraft wing.

This brings us to another unique quality of a bird, which sets it apart from almost any other creature : Its respiratory system. Have you ever wondered how birds can breathe very thin air at 8,000 metres or more ?

If we were at that height in the open air, we'd risk serious illness due to lack of oxygen. So, how do they do it ?

Another thing : Birds are able to fly for hours, even days beating their wings. This creates considerable heat which leads to another question how do they cool off since they don't have sweat glands ?

In answer to both these questions, it all comes down to a special system of air sacs in just about all the important parts of their body. These air sacs help the internal circulation that keeps them cool. Unlike reptiles whose lungs take in air and expel it, in birds there is a constant flow of air whether they are inhaling or exhaling. It works like this : The air the bird inhales goes to air sacs which then push air into the lungs like bellows.

From there it goes into other air sacs and is eventually expelled, which means there is always fresh air passing through the lungs.

The blood in the lungs flows in the opposite direction, and its this counter-current between air and blood that enables the bird to breathe the thin air of high altitudes. This method of extracting oxygen from air is the most efficient in the animal kingdom. Far superior to the respiratory system of a reptile.

Egglaying is another difference. Birds incubate their eggs, and most have a brood spot on their breast, devoid of feathers, designed for this purpose, with a network of blood vessels to keep the eggs warm. Interestingly, the few birds without this spot, instinctively make a brood spot by pulling out the feathers in that area.

Regarding this point of instinct peculiar to birds, they would need completely new ones if they evolved from reptiles. For example : nest building, hatching eggs, and feeding the young. Owing to the fact that instinct cannot be taught but only passed to the young genetically, how did the birds instincts get passed on from reptiles ?

Another outstanding feature of a bird as opposed to a reptile is their eyesight. We all know of the legendary eyes of an eagle and a hawk, but even little warblers have eyes like telescopes. There are more sensory cells in birds eyes than any other living thing. Think of the changes needed to evolve from a reptile.

Other changes needed would be their feet. Birds have only four toes, reptiles have five. Unlike the reptile, birds have no vocal chords. All those melodious songs they make come from a thing called a syrinx. Their heart is different too. A bird has a four-chambered heart, a reptile three.

Having considered all the facts, Le Comte du Nouy made the admission that birds have "all the unsatisfactory characteristics of absolute creation".   (15)

Because of the difficulties of linking the reptile with birds, many scientists believe the reptile might have evolved into a mammal thereby circumventing the bird/reptile dilemma. Unfortunately, the problem is almost as pronounced as the reptile into bird saga.

The very fact they are called "mammals" points to a major difference to the reptile: That of having mammary glands which provide milk for their young, which are born alive. The mammalian young instinctively know they need to suck milk from their mother, and have the muscles to do this.

This is quite different from baby reptiles who lack this instinct or ability. For example a baby crocodile, after hatching from the egg usually fends for itself, instinctively looking for food, independent of its mother.

Another feature found in a mammal which is absent in a reptile, is the placenta which is a highly complex means of nourishment for the unborn young. Reptiles lack this. They also lack a diaphragm found in mammals.

Mammals have three bones in their ears, reptiles just one. Mammals have constant body temperature as a rule, reptiles don't. Regarding bone structure, reptilian legs are fixed at the side of their body causing their belly to be close to the ground whereas mammals legs are underneath the body which raises it above the ground.

Another thing that sets them apart was pointed out by evolutionist. Theodosius Dobzhansky: "Mammals have greatly elaborated their teeth. Instead of the simple peg-like teeth of the reptile, there is a great variety of mammalian teeth adapted for nipping, grasping, piercing, cutting, pounding, or grinding food". (16)

So, what do you think so far? Do evolutionists have all the facts or are their ideas a fallacy? You decide.

If there are gaps that exist between the preceding creatures that are hard to bridge, they are as nothing compared to the gulf that exists between mammals and man, as we shall see in the next chapter.

CHAPTER 5

Some, but by no means all evolutionists believe that apes were our ancestors. George Gaylord Simpson was one who felt this way as we can see from this quote : "The common ancestor would certainly be called an ape or a monkey in popular speech by anybody who saw it. Since the term APE and MONKEY are defined by popular usage, mans ancestors WERE apes or monkeys". (17)

Is there any evidence to support this assertion ? Consider this: If the so-called APE-MAN was supposedly much more advanced in the evolutionary process than apes and monkeys, how come they died out leaving no real evidence they existed while the inferior apes and monkeys are alive and well ? Why would the links that were higher up the chain become extinct, while the lower forms survive? Somethings wrong here.

But what about all those bones dug up all over the place that are said to belong to a hominid or ape-man ?

It may surprise you just how sparse the "evidence" is. SCIENCE DIGEST makes this comment : "The remarkable fact is that all the physical evidence we have for human evolution can still be placed, with room to spare, inside a single coffin ! ....Modern apes for instance, seem to have sprung from nowhere. They have no yesterday, no fossil record. And the true origin of modern humans – of upright, naked, tool-making, big brained beings — is, if we are to be honest with ourselves, an equally mysterious matter". (18)

But what about all those drawings we see in scientific textbooks? Are they not based on solid evidence? SCIENCE DIGEST gives the answer : "The vast majority of artists conceptions are based more on imagination than on evidence......Artists must create something between an ape and a human being; the older the specimen is said to be, the more ape-like they make it".

Another blow to the ape into man theory is that primitive man, although living occasionally in caves, was not as primitive as first thought. They were able to do things that were completely alien to a brute beast such as an ape.

In the book MAN, GOD and MAGIC, Ivar Lissner commented : "Just as we are slowly learning that primitive men are not necessarily savages, so we must learn to realize that the early men of the ice-age were neither brute beasts nor semi-apes, nor cretins. Hence the ineffable stupidity of all attempts to reconstruct Neanderthal or Peking man".   (19)

Surely there must be plenty of evidence in the form of skeletons of monkeys and ape-men you might ask ? Decide for yourself from this comment from SCIENTIFIC AMERICAN : "Primatologists may therefore be forgiven their fumblings over great gaps of millions of years from which we do not possess a single complete monkey skeleton, let alone the skeleton of a human forerunner..........we have to read the story of primate evolution from a few handfuls of broken bones and teeth. These fossils moreover, are from places thousands of miles apart in the old world.......In the end we shake our heads, baffled......It is as though we stood at the heart of a maze and no longer remembered how we had come there".

If the scientific investigators from CSI have difficulty establishing a victims age or sex from a handful of bones of recent vintage, how much harder it must be to create an ape-man from "a few handfuls of bones and teeth" thousands of years old ?

What about all those charts depicting ape-like creatures with weird names, reputed to be our common ancestors ? The first specimen on the chart has the name Propliopithecus. It was found in Egypt and closely resembles a gibbon. The next on the chart is called Dryopithecus and was found in Africa. There are huge gaps between these finds of as much as nineteen million years. Rather than claim these as our common ancestors some evolutionists are putting the Propliopithecus in the line that leads up to gibbons, and the Dryopithecus in another line leading to apes. Man is not in the loop at all !

Perhaps the most well known alleged ancestor of man is called Australopithecus. This creature is believed to have been a tool-maker, although the size of its brain was about a third that of modern man.

Is there scientific evidence that they were really like us? Evolutionist Le Gros Clark furnishes the answer: "The terms 'man' and 'human' can only be applied to them with some reserve, for there is no certain evidence that they possessed any of the special attributes which are commonly associated with the human beings of today".

Some years ago a discovery was made in Sterkfontein in South Africa of 58 stone artifacts. Because the area where they were found also had the remains of Australopithecines it was assumed the latter were toolmakers. But does this fit the facts ? Not according to SCIENCE magazine which made this comment : "Mason thinks that tool-making of the complexity shown in the Sterkfontein industry was probably beyond the ability of the Australopithecines and that it must be ascribed to some more advanced hominid.

Ashley Montagu sums it all up in his book   MAN, HIS FIRST MILLION YEARS when he said :   "The skull form of all Australopithicenes is extremely ape-like........such creatures could not have been directly ancestral to man...........the Australopithecines show too many specialized and ape-like characters to be either direct ancestors to man or of the line that led to man".

But what about the famous Neanderthal man found in Germany ? Is he the vital missing link between ape and man ?

Interestingly, TIME magazine (March 19th 1961 issue) referred to the brain capacity of Neanderthal man as being 1,625cc, which is actually larger than that of the average modern man.

WORLD BOOK ENCYLOPAEDIA gives the following description of this ancient man :   "At first, scientists thought that Neanderthal man was a squat, stooping, brutish, somewhat apelike creature.   But later research showed that the bodies of Neanderthal men and women were completely human, fully erect, and very muscular.   There brains were as large as those of modern man".   (20)

Another problem for evolution believers is the identification of skulls. If an ancient skull is discovered it is extremely difficult if not impossible to to even identify which racial group it belongs to. For example, the racial types with the most distinctive shaped skulls of any humans, are those of negroes and Eskimo's. Now if one each of these skulls were placed side by side, and experts were brought in to establish which skull belonged to what race, how easy do you think it would be ?

For the answer, note this observation by Le Gros Clark in the book : THE FOSSIL EVIDENCE FOR HUMAN EVOLUTION : "Now it is probable that that there are no racial types in which the skull characters are more distinctive than negroes and Eskimo's; and yet experts fail to agree when faced with single skulls whose claims to these types are in question. If a decision proves so difficult in such cases, it will be realized how much more difficult , or even impossible it will be to identify by reference to limited skeletal remains, minor racial groups with less distinctive characters".

Some evolutionists have been so desperate to prove their theory, they have resorted to fraud to support their ideas. The most famous case is that of the so-called Piltdown man "discovered" in England. Circumstantial evidence points to a Mr Charles Dawson as being the perpetrator of the fraud. Apparantly, what he did was to take the skull of a modern man, staining it to make it look older than it really was, and filing down the teeth to make them appear worn. Then, taking the jawbone of an ape he doctored it up with bichromate of potash and iron to make it look mineralized.

With reference to the fraud, READERS DIGEST wrote : "Every important piece proved a forgery. Piltdown man was a fraud from start to finish! ....all the circumstantial evidence points to Dawson as the author of the hoax".

The disturbing thing about this saga, is that it was touted as firm evidence to support evolution FOR OVER 40 YEARS ! This sort of fraud was not the first and by no means the last. This disgraceful behaviour should really make you think twice when some scientists promote some new "discovery". Be alert, be skeptical, be distrusting and careful. They deserve no less.

A major difference in a chimpanzee or ape over a human is the size of their brain. Most of our thought, intelligence, motivation and even our personality, comes from the section of the brain known as the prefrontal cortex. With it we are able to process abstract ideas, make plans, judgements and exercise our conscience. Whereas our prefrontal cortex is quite large, in animals this area is either small or nonexistent. One scientific journal made the point that if a humans cerebral cortex were flattened out it would cover four pages of typing paper, whereas a chimps would only cover one page and a rat's a postage stamp.

Something else that sets us apart from apes is our use of language. It truly is a marvellous gift. Whenever we talk we use some hundred muscles in the tongue, lips, jaw, throat and chest. How it works is well explained by speech expert Dr Wilkins H Perkins when he said : "We utter about fourteen sounds a second. Thats twice as fast as we control our tongue, lips, jaw or any other parts of our speech mechanism when we move them separately. But put them altogether for speech and they work the way fingers of expert typists and concert pianists do. Their movements overlap in a symphony of exquisite timing".

A part of our body that we often take for granted is our hands. A human hand is a remarkably versatile tool. With its opposable thumb we're able to thread a needle, hammer in a nail, pick up a phone, write a letter and do a host of other things. Are we not in awe of a concert pianist and the dexterity of his fingers as they fly over the keys, producing melodious music ? Imagine the coordination needed between his brain and his fingers.

This brings us to another difference when it comes to human versus animal. Professor Guyton's TEXTBOOK OF MEDICAL PHYSIOLOGY when discussing the human motor cortex, (the part of the brain that tells the body what to do) says concerning it : "Is quite different from that of lower animals" making possible "an exceptional capability to use the hand, the fingers, and the thumb to perform highly dexterous manual tasks".

Compare this to the hands of an ape : "The apes, having short thumbs and long fingers, are handicapped in relation to delicate manual dexterity". (THE NEW ENCYCLOPAEDIA BRIT.). Their long fingers are ideal for swinging from limb to limb up in the trees, but threading a needle or playing the piano ? I don't think so.

Concerning the hand, I think it would be appropriate to quote one of the greatest scientists who ever lived......Sir Isaac Newton who said : "In the absence of any other proof, the thumb alone would convince me of God's existence".

One final gap between man and monkey we will consider is that of DNA.

It was long thought by scientists that the genetic makeup of a human and an ape were similar. In recent times they've had to revise their findings.

"Large differences in DNA, not small ones, separate apes and monkeys from both humans and each other".

Kelly Frazer who worked for the company who did the above analysis explained it this way : "There are large deletions and insertions sprinkled throughout the chromosome." Summing up the differences, the above article said : "There is a yawning gap that divides monkeys and us".

Because the DNA is such a vital factor in deciding the shape and character of living things on earth, the next chapter will deal with this subject.

CHAPTER 6

Under a powerful microscope the DNA has the appearance of a twisted ladder. It is made up of two strands wound around each other, connected by combinations of four compounds called bases. The base of one strand is paired with a base on the other strand forming the rungs of the DNA ladder. What determines the kind of information carried is the exact order of the bases in the DNA molecule. It is this sequence that decides our physical appearance; the shape of our nose, ears and mouth, and the colour of our hair, eyes and skin.

Proteins, which are made up of amino acids are perhaps the most important compound in the living cell, accounting for nearly half the dry weight of most organisms. Some are made by your body and others from your diet.

Their functions are many. Take for example hemoglobin. This found in red blood cells and its job is to transport oxygen throughout the body. The function of antibodies is to help your body ward of disease.

Insulin which is also a protein help your body metabolize food. A single cell may contain hundreds of proteins all carrying out specific jobs.

Remember that all this activity is going on within a cell so tiny that five hundred of them could fit inside the period at the end of this sentence.

Paul Davies, in his book THE FIFTH MIRACLE describes just how intricate the cell really is. He states : "Each cell is packed with tiny structures that might have come from an engineers manual. Minuscule tweezers, scissors, pumps, levers, valves, pipes, chains, and even vehicles abound. But of course the cell is more than just a bag of gadgets. The various components fit together to form a smoothly functioning whole, like an elaborate factory production line".

Whether a plant or a human we all start out as a single cell. This then divides into two cells, which divide into four and so on. At a certain point the cells begin to specialize, some becoming nerve cells, muscle or skin cells. In the ongoing process cells will group together to form tissues. Some will eventually form organs such as the heart, lungs and eyes.

The nucleus of the cell is considered to be the control centre as it directs nearly all the cells activities. Its within this nucleus that the genetic program lies, written out in deoxyribonucleic acid, or DNA for short.

The molecules of the DNA lie tightly coiled in what we call chromosomes. The sections of DNA molecules which contain all the information that makes us what we are, are called genes. This is what makes us unique, and enables criminal investigators to track down a murderer from a tiny speck of DNA.

As important as DNA is, on its own it cannot function properly without the help of an intermediary called ribonucleic acid (RNA). This has the job of transferring the DNA code out of the cells nucleus so it can be decoded by things called ribosomes. Ribosomes are in fact protein producing factories. The information is also "read" by other specialized proteins.

Before cells divide they must replicate the DNA. How do they do this ?

Imagine the 'twisted ladder' as a zipper. Proteins come along and unzip sections of the double stranded DNA. Then following strict roles of pairing free bases in the cell are linked together with their matching bases on the original strands. Once the pairing is complete we're left with two duplicate codebooks. So now, when the cell divides, each new cell gets an identical DNA codebook.

Now to the next step in this amazing process : The making of proteins.

First a special protein zips open a section of the DNA strands, then RNA bases that are roaming free within the cell link up with the now exposed DNA bases on one strand only. This now becomes a strand of messenger RNA. This now peels off and moves away toward the ribosomes. A transfer RNA (which closely resembles R2D2 the little robot from the Star Wars movies) picks up an amino acid and brings it to the ribosome which sweeps across the messenger RNA, resulting in a chain of amino acids, all linked together. While forming, the chain starts to fold into the shape needed to function properly. Once this is complete, it is released by the ribosome.

All this complicated arrangement within the cell creates a major dilemma for the evolutionist. The very word 'evolution' implies a gradual forming of proteins into a living cell. DNA alone cannot create life as it requires the aid of RNA and many other specialized proteins. Also let us not forget the role of the ribosome, which is basically a factory. All these components, to create life, would have to all work together at the same time, they can't work independently of each other. Therefore a GRADUAL process would really be impossible.

With all the amazing knowledge now acquired of genetics and molecular biology, one thing stands out above all else : LIFE DEPENDS ON HAVING ALL THESE ELEMENTS SIMULTANEOUSLY ! It could never have come about by chance, spontaneously; it all had to work from the very beginning. It simply does not allow for a gradual evolving of all the separate parts that make up the cell. For it to work, the DNA and RNA, along with the ribosomes and many specialized proteins all had to exist at the one time and in the one place.

What are the chances of atoms collecting together to form the simplest self producing cell ? In his book A GUIDED TOUR OF THE LIVING CELL, Nobel prize-winning scientist Christian de Duve admits : "If you equate the probability of the birth of a bacterial cell to that of the chance assembly of its component atoms, even eternity will not suffice to produce one for you".

In view of the preceding facts, does it not prove that the living cell does not have a simple beginning as is the wish of evolutionists ? And the fact that the cell is the basic building block of all life on earth, and scientifically could not come about by accident, does this not show, at the very outset, evolutionists haven't a leg to stand on.

The DNA constitutes a built-in code, a blueprint that keeps all life forms within their basic kinds. It makes no allowance for alterations, unless something from outside such as radiation occurs. If this happens, it never improves the organism, but only impairs its progress.

The DNA allows for incredible variety among humans and animals. It has been said that no two humans are exactly alike, and the same could apply in the animal world. Having said that humans and animals are basically made up of the same chemicals and elements. As science writer Rutherford Platt put it : "These DNA specks have a similar chemical composition, are about the same size, and look very much like those in your dog, or in a house fly, a bread mold or blade of grass. Yet somehow the specks are coded to make every living thing different from every other living thing. They make dogs different from fish and birds, bread mold from apple trees, elephants from mosquitoes".

Because all organisms are made up of the same basic elements, take in nourishments from those same elements, live on the same planet and are under the same physical laws, its not surprising there is similarity of design. But there are of course differences that allow some to operate in the air, others on land, and still others in the water. The DNA instructions decide peculiarities of all living things. It is this heredity that keeps family kinds separate creating a wide gulf between them.

Both scientifically and realistically; at the very beginning of life on earth, there had to be DNA already in place, issuing instructions to a particular cell, as to how it would grow, whether into a tiny fish, a plant or whatever.

Not only did the DNA have to exist from day one it also required RNA to release and transport the information, and the ribosomes to decode it. Also, to complicate things further, many specialized proteins had to exist at the same time and in the right sequence. This scientific fact makes the evolution of life impossible.

Our discussion of the complexities of DNA and the world of genetics leads us to another closely related subject : The fixity of all the various species on earth.

CHAPTER 7

SCIENTIFIC AMERICAN made an interesting comment concerning a fundamental law of all living things. It said : "Living things are enormously diverse in form, but form is remarkably constant within any given line of decent : pigs remain pigs, and oak trees remain oak trees, generation after generation". (21)

This basic fact is an unchangeable scientific law on earth. The science of nomenclature is the system used to establish which animal and plant groupings are related. Individual organisms that are very closely related are considered SPECIES. One or more of these make up a GENUS. A group of these (genera) are called a family. As an example, consider the cat family, FELIDAE. One genus of this family is FELIS, which includes the Tiger, Lion, house cat and some others. These are all separate species within the genus.

We all know that certain members of the cat family can mate and produce offspring even though they are different in appearance. For example a tiger and lion have mated and produced a creature called a Liger. But a lion and tiger cannot mate with a house cat or bobcat because of size and other reasons. So therefore, even within the same family, there are limitations. The basic fact remains that within the group known as Felidae or cat family, cats will always remain cats and could never cross over and mate with another family group such as a dog.

Within the dog family there is a great variety, such as domestic dogs, wolves, dingoes, foxes and so on. But they can only mate within their group.

Even though there is huge diversity in the animal, fish and bird world within their family kinds, its a scientific fact that the separate kinds cannot mix, no matter what.

There is not a single shred of evidence to indicate that the basic kinds ever evolved from a common ancestor. The main reason for this is because the germ cells of one kind cannot unite with the germ cells of another kind they are incompatible, making it impossible. Hence the amoeba stayed forever an amoeba, a fly stayed forever a fly, and an ape forever an ape.

One interesting aspect of reproduction within species is the occasional result of a hybrid. At one time evolutionists got very excited about this, thinking it might be proof that organisms could make the crossing over to a different kind. But does it happen that way ? Are there insurmountable obstacles ?

Just consider the horse for a moment. If an average male horse was left alone in a field with a female donkey, and the inevitable happened, the result would eventually be a mule which is a hybrid. Now herein lies the problem for the evolutionist : The mule is sterile and cannot reproduce. It has reached what scientists call the BARRIER OF STERILITY.

Because humans are all the same 'kind' (mankind) we don't have a problem. Thus all varieties of humans are fertile with one another. The tallest and the smallest can mate and have offspring. For example a pygmy can mate with a very tall Scandinavian without any problem.

With animals and plants, any hybrids that result come from things that are closely related to begin with. Many of these hybrids are sterile and in the free state usually don't breed. For those that are fertile, further hybridizing reaches a final limit which it cannot go beyond.

This limitation is well illustrated with corn. For decades agricultural scientists have made amazing progress in raising high-yield hybrid corn.

Then it reached a point where the hybrid corn seed could no longer be improved because all the factors for improving useful characteristics had been utilized. It had reached its limit of improvement. Besides, no matter what they did to the corn it still remained corn and never changed into another form of plant life.

Ah, say the evolutionists, what about the adaptability of plants and animals. Isn't this proof of evolution ? No it isn't, because plants and animals have built-in mechanisms to adapt to a greater or lesser degree.

Some point to the polar bear as an example of a creature evolving into a cold weather animal. Obviously the common brown bear could have adapted to the harsh climate of arctic regions. But the point is : The Polar bear can survive quite well in a warmer climate, as we know there are many of them spread throughout zoos all over the world. Is it possible that it had the ability to adapt to cold weather better than other animals ?

Regarding this ability to adapt to new surroundings, evolutionist Dobzhansky makes this observation : "The English sparrow introduced in the United States from Europe has changed detectably in its new home; the average size of the birds has increased, and they became differentiated into incipient local races".

What does the above prove ? That the potential for change was already present. The sparrow continues to be a sparrow, only larger. It could never change into another specie and never will. Quite often, adaptability is confused with evolution , and they are definitely not the same.

Still on the subject of birds, some evolutionists point to the woodpecker as an example of a bird evolving certain advantages for survival. Note the argument used by De Beer in his book : "The woodpecker has two toes on each foot pointing backward, enabling it to get a firm foothold on the bark of trees, stiff tail feathers serving to prop it securely against the tree, a long stout beak with which it chisels holes in the bark, and a very long tongue with which it reaches and takes the grubs at the bottom of the holes........these adaptions must have arisen during the evolution of the woodpecker".

The question remains, what was the woodpecker doing to survive before it got its backward toes, long beak and tongue ? If it managed to survive with a different foot, shorter beak and tongue for some while, why develop these other features ? And if these were necessary for survival, how did all the other birds with shorter beaks survive ? It just goes to prove that when it comes to food gathering, birds of all kinds gather food in different ways and had certain characteristics and could also adjust to a changing environment.

If different feeding habits are attributed to evolution, making it easier for some to survive than others, then what about the cow and the horse ?

Both eat the same grass in the same meadow. Why is it one evolved upper front teeth and the other didn't. How could both exist side by side, each better suited for survival , one because it had those teeth and the other because it didn't ?

Another example of adaptability which is sometimes confused with evolution is what occurred some years back with flies. When many were exposed to the insecticide DDT, most were killed on contact. But some were able to resist the DDT. They survived and produced offspring that were also resistant, but they still remained flies.

What this meant was, some flies had greater resistance or adaptability than others, but this was not due to evolution as some have claimed.

This business of changing brings us to another subject which evolutionists seize on : That of mutations. We will discuss this in the next chapter.

CHAPTER 8

The very foundation of the theory of evolution is built on the belief that small mutations over time caused changes that brought a slow transition of one form of life into another. Before we examine the facts, lets look at the meaning of mutation.

The word is taken from the Latin MUTARE meaning 'to change'. What this means is that when an alteration of some kind happens within the germ plasm of a cell, this change is called a mutation and is inheritable.

Familiar examples of what mutations are capable of, are people who inherit Albinism and Downs Syndrome. Both have a physical appearance which makes them different from the majority. For instance, an Albino has white hair and skin, and usually pink eyes due to a congenital absence of pigment.

Why do mutations occur ? BIOLOGY FOR TODAY gives this answer : "Mutations probably occur due to factors normally found in the environment : cosmic rays and other ionizing radiations; metabolic processes in cells; or errors in gene replication".

The above mentioned example of the Albino would fall into the category of errors in gene replication. A similar situation would apply to a person with Downs Syndrome.

A striking example of what radiation can do to a human, is what occurred in Hiroshima and Nagasaki, Japan in 1945. The two atomic bombs dropped on those cities caused mutations on a grand scale. For those unfortunates that survived, many were horribly deformed as a result.

Another cause of mutations are chemicals. Perhaps the most well-known case of the disaster they can cause is that of the tranquilizer drug thalidomide. What a terrible shock to parents to see their newborn baby without arms and legs or some other horrible deformity.

Waste chemicals that end up in rivers have been known to cause horrific mutations on fish. For example some have developed one single eye, and others have been hatched with two heads.

Scientists have experimented on animals using different chemicals to see what will result. They've come up with featherless chickens; insects having eyes of an abnormal colour; and changes in the size of wings and limbs of other creatures.

Some have hailed mutations as the answer to the origin of the species. They say that over aeons of time and millions of mutations, we end up with the life forms of today. But does this fit the facts ? Consider the foregoing examples. Would a featherless chicken have any advantage over one WITH feathers ? Certainly not in winter. Would a fish with two heads or one eye have any advantage over normal healthy fish ? Experts say no, claiming the mutation has made them inferior and poorly suited for survival.

Would a child born without limbs have any advantage over a normal child ? The answer is obvious.

The one constant that applies to almost all mutations is that they are degenerative. This fact was reported in SCIENTIFIC AMERICAN when it said : "In more than 99 percent of cases the mutation of a gene produces some kind of harmful affect, some disturbance of function".

Think about it. If ninety nine out of a hundred mutations are harmful, making the recipient of the mutation less suited to survival, how could any living thing get the chance for improvement ?

Mutations are often compared to accidents in the genetic machinery of living things. Its akin to the wrecking of a vehicle. If we were to grant that one out of a hundred accidents MIGHT improve the vehicle, what of the ninety nine that are harmful. How much would be left after ninety nine harmful accidents ? In any case it still remains a vehicle, it hasn't changed into something else, the only difference being its now inferior.

Similarly with mutations of genes. Not only are they inferior to normal healthy ones, but they do not evolve into something else. If the mutated gene was in a pig, it still remains a pig and not some other animal.

Ah say the evolutionists, you have to allow billions of years for beneficial changes to happen and that could make the difference. But does it? Consider this : If you did not maintain your car, what would happen to it over lots of time? Wouldn't it just rust away to nothing?

Take the crashing of waves on the seashore. Over much time they can erode huge cliffs of rock. Time is DE structive not CON structive.

Of all the experiments made by scientists with mutations NOT ONE has ever resulted in a new KIND of animal or plant. Whatever changes take place, the mutated specie always remains the same. This is a proven scientific fact.

Evolutionist Dobzhansky made this admission : "A majority of mutations, both those arising in laboratories and those stored in natural populations, produce deteriorations of the viability, hereditary diseases, and monstrosities. Such changes, it would seem, can hardly serve as evolutionary building blocks".

As mentioned at the outset of this chapter, some believe that mutations form the foundation of their theory, yet as one of their number just quoted admits that mutations "can hardly serve as evolutionary building blocks".

In summary, what do we find? We discover that the evolution theory has no foundation to build on, and no building blocks to build with. Hardly does much to build your confidence in it, does it?

Probably by this time, you the reader may be wondering : What facts ARE THERE to support evolution? Surely there must be evidence in the fossil record you say. This we will consider in the next chapter.

CHAPTER 9

What exactly are fossils ? They are the remains of ancient life forms preserved in the earths crust. Many fossils no longer contain the original material but are made up of mineral deposits that have infiltrated them and taken on their shape. So you would have the remains say of a fish, that looks exactly like what it is, but is made up of sandstone or rock. The imprint of it being made under great pressure some millenniums ago.

It has long been the hope of evolutionists, that the fossil record would reveal a gradual changing from one kind of life into another. Has it proved supportive of their theory ?

Note what Darwin himself concedes : "The distinctiveness of specific life forms and their not being blended together by innumerable transitional links, is a very obvious difficulty". (22)

In laymen terms what Darwin is saying is that because the discoveries in fossils are so distinctive, that a fish is immediately recognizable as in all other fossils, with no signs of a transitional link to some other form of life it creates a major problem for the theory.

The fossil record has always been important to evolutionists because it was their one hope of finding the beginning of new structures in living things. If they could just find some fossils with developing arms, legs, wings, eyes and other bones and organs, things like fish fins turning into amphibian legs, and gills into lungs, then they would have absolute proof of evolution. And indeed they would if that was the case. But what are the facts ?

Does the fossil record reveal a gradual change taking place within organisms ?

Let Darwin himself answer : "The abrupt manner in which whole groups of species suddenly appear in certain formations has been urged by several paleontologists.......as a fatal objection to the belief in transmutation of species". (23)

One way in which Darwin tried to overcome this problem was to describe the fossil record as "imperfect" or incomplete. He claimed that there was not enough fossils available to prove the point one way or another. Its true that in his day the accumulation of fossils was quite sparse. But what of the ensuing one and a half centuries. In that time there has been extensive digging all over the world with vast numbers being unearthed.

Porter Kier, a scientist with the Smithsonian Institution had this to say : "There are a hundred million fossils, all catalogued and identified in museums around the world." (24)

The thing that has confounded such scientists as the above is the fact that all this massive amount of evidence now available, reveals the very same thing it did in Darwin's day : That basic kinds appeared suddenly with no appreciable changes over a very long period of time. No transitional links have ever been found in the fossil record.

After forty years of research, Swedish Botanist Heribert Nilsson expressed his frustration in this way : "It is not even possible to make a caricature of an evolution out of palaeobiological facts. The fossil material is now so complete that.......the lack of transitional series cannot be explained due to the scarcity of material. The deficiencies are real, they will never be filled". (25)

At the beginning of earths existence, scientists believed there were many single celled organisms that eventually multiplied and turned into many-celled ones. They believed that the first cells were simple and evolved into complex cells. What do the facts show ? Is the fossil record of the earliest times helpful ?

In his book RED GIANTS AND WHITE DWARFS Robert Jastrow makes this observation : "The critical first billion years during which life began are blank pages in the earths history". (26)

All the fossil residues found in later times in rocks, do not reveal a simple beginning, quite the opposite. To quote the book

EVOLUTION FROM SPACE : "Most of the biochemical complexity of life was present already at the time the oldest surface rocks of the earth were formed". (27)

When what scientists call the Cambrian period began, the fossil record takes an unexplained dramatic turn. There is a sudden explosion of living things; complex sea creatures, many with hard outer shells. There are snails, sponges, starfish and lobster-like things called trilobites.

Are there are any links to this outburst of life and the much earlier period ? Listen to the words of Darwin on this subject : "To the question why we do not find fossiliferous deposits belonging to these assumed earliest periods prior to the Cambrian system, I can give no satisfactory answer". (28)

What about comparisons between fossilized creatures or plants with their modern counterparts ? Do we find a big difference in their appearance ? Well, take as an example insects which have been found in large numbers. A fossil fly which had been labelled "40 million years old" was examined closely by a scientist , a Dr. George Poinar Jnr.

He said : "The internal anatomy of these creatures is remarkably similar to what you find in flies today. The wings and legs and head, and even the cells inside, are very modern looking". (29)

Even when you go back much further in the ancient fossil record you find the same thing. For example DISCOVER magazine says : "The horseshoe crab.......has existed on earth virtually unchanged for 200 million years". (30)

The same similarity between ancient and modern, can be found in the plant world. Fossils have been found of many trees and shrubs that show very little difference to the leaves of today. Oak, walnut, hickory, grape,magnolia, and palm are instantly recognizable to a person experienced in horticulture.

An extensive study was carried out by the Geological Society of London involving 120 scientists, all specialists. The final result was a monumental work of 800 pages, presenting the fossil record for plants and animals divided into about 2,500 groups. A report of this study went on to say:

"Each major form or kind of plant and animal is shown to have a separate and distinct history from all other forms or kinds.

Groups of both plants and animals APPEAR SUDDENLY in the fossil record......Whales, bats, horses, primates, elephants, hares, squirrels etc, all are as distinct at their first appearance as they are now. There is not a trace of a common ancestor, much less a link with any reptile, the supposed progenitor".   (31)

Much to the chagrin of evolutionists, the fossil records testimony is creation oriented. It shows that the many different kinds of living things appeared suddenly. This is what you'd expect if these had been created.

While there is huge variety within each species, the fossil record shows absolutely no link to any evolutionary ancestor before them.

To sum up this chapter, I could do no better than quote the words of zoologist Harold Coffin : "To secular scientists, the fossils, evidences of life in the past, constitute the ultimate and final court of appeal, because the fossil record is the only authentic history of life available to science.

If this fossil history does not agree with evolutionary theory — what does it teach ? It tells us that plants and animals were created in their basic forms. The basic facts of the fossil record support creation, not evolution". (32)

As can be seen, evidence thrown up by the fossil record is of no help to believers in evolution, which may come as a surprise to many who always thought there was plenty of evidence to support it.

CHAPTER 10

For this chapter, it would be good to review the different claims of evolutionists and the facts as they stand.

For example, at the very beginning of earths existence evolutionists speculated about the type of atmosphere that was around. They believed it was composed of carbon dioxide, methane, ammonia, and water. Energy from either the sun or lightening broke up these compounds, forming them into amino acids. They believed that somehow at some point in time these acids came together and made a protein, the basic building block of life. 20 amino acids are required to produce a protein.

After decades of experiments made in the laboratory scientists have not been able to reproduce life in a test-tube. They always fell short of the twenty necessary amino acids.

One of the problems they faced , is when they introduced a spark to break up the chemicals in the 'atmosphere' it quickly destroyed the amino acids.

As they were simulating the effect of lightening on the acids, it became clear that at the beginning, no amino acid could survive a bolt of lightening if it couldn't survive a mere spark in a test-tube.

But what about the suns rays, would they have been more conducive to creating a protein ? The facts say no. Because emanating from the sun are harmful ultraviolet rays and we all know what they're capable of.

The evolutionists try to get around this dilemma by saying the amino acids must have plunged beneath the ocean for protection, which would have been a good argument but for one thing : Water will destabalize amino acids inhibiting their growth. But the greatest dilemma of all is the simple fact : "With oxygen in the air, the first amino acids would never have got started; without oxygen it would have been wiped out by cosmic rays".

If we were to ignore the preceding facts and assume that SOMEHOW the amino acids survived the lightening/sun/water and oxygen, what were the chances of them randomly forming into a protein ? We learned in the third chapter that there are over a hundred amino acids in existence, but only twenty are needed for life. We now know that these twenty are all left-handed. The scientists are baffled as to the reason, but it sure increases the odds of them all uniting at random.

To illustrate the impossibility of this, imagine you have a pile of beans the kind you put in soups, say Lima and Kidney beans. Lima are white in colour and Kidney red. You then plunge a scoop into the pile of totally mixed beans. When you withdraw the scoop what do you think your chances are that they'd all be red ? Nil ! It would be absolutely impossible. So it goes with the random separation of twenty amino acids from the hundred or so that exist.

The mathematical odds against this happening is one followed by 113 zeros. Mathematicians dismiss this as never happening, simply because the above number is larger than the estimated number of ALL THE ATOMS IN THE ENTIRE UNIVERSE !

A further difficulty is that old chestnut, "which came first, the chicken or the egg"? This is mentioned because proteins cannot be assembled without the nucleic acids DNA and RNA. But nucleic acids can't form without proteins. Hence that old riddle which came first ? This is a problem of gigantic proportions, which evolutionists have not come up with a satisfactory answer.

Still on the subject of lifes origins, in chapter 3 we considered the dream of all evolutionists of finding a living cell which was very basic or simple. What do the facts show ? That no such cell exists. Bacterial cells are believed to be the simplest of all living systems, and what do super microscopes reveal ? They are exceedingly complex with "exquisitely designed pieces of intricate molecular machinery........far more complicated than any machine built by man and absolutely without parallel in the non-living world".

In the following chapter, we examined the first major step in the evolution theory, that of from fish to amphibian. It is believed that somehow fish fins turned into jointed limbs, and gills into lungs, like that of a frog. But alas, it didn't fit the facts. Fish have a two chambered heart whereas a frog has one with three chambers. Fish absorb sounds through their bodies, frogs have eardrums. Their backbones are so different they would need to be changed beyond all recognition.

The next step in the 'chain' was from amphibian to reptile. A major difference here is the fact that the eggs of amphibians are fertilized EXTERNALLY whereas the egg of the reptile is fertilized INTERNALLY before the shell is formed. Disposal of wastes is another difference.

Amphibian embryos release their wastes in the surrounding water. With reptiles the wastes are stored within the egg in a special membrane for that very purpose. As you can see, these are quite serious differences.

Going on to the next step in the assumed evolutionary process, the reptile is supposed to turn into a bird. Scales are thought to evolve into feathers. How scientific is this idea ? In chapter 4 we learned about the amazing design of the feather. Along its shaft there are things called barbs. Each barb is connected by barbules which in turn has hundreds of little barbicels and hooklets. Altogether it has no equal as an airfoil.

When scientists examined a pigeon feather under a microscope they discovered several hundred thousand barbules and MILLIONS of barbicels and hooklets. In your honest opinion, do you really think there could ever be a connection between a scale of a reptile and the feather of a bird ?

Another thing we learned earlier is the difference of the density of the bones of these two creatures. To be able to fly, the bird would naturally need very light bones, albeit strong. For this reason their bones are thin and hollow with struts inside for strength. Totally different to the solid bones of a reptile. Big change needed here.

Other differences : Respiratory system, blood type, (warm as opposed to cold), vocal chords, (birds don't have them, reptiles do), reptiles have four toes, birds have three, a bird has a four chambered heart, a reptile three. The bird has a beak, the reptile a nose. Really, there are so many disparities between the two its led evolutionists to try and circumvent the reptile into bird problem by saying the reptile must have evolved into a mammal.

Unfortunately for them, the problems are almost as acute as the reptile into bird saga. For example, the very fact they are called "mammals" indicates a major difference with the reptile. A mammal has special glands which provide milk for their young which are born alive. Another disparity is the placenta, which is a vital part of a mammal, completely absent in a reptile. Mammals have a fairly constant body temperature, reptiles don't. In their ears, mammals have three bones, reptiles one.

Mammals legs are underneath the body, reptiles are fixed at the side of their body. Mammals have elaborate teeth, reptiles simple peglike teeth. And so it goes on.

Now we come to the most well known part of the evolution theory, that of ape into man. What were some of the facts brought out earlier ?

FACT (1) If so-called ape-man was the intermediary between apes and man, how come the "superior" ape-man disappeared without a trace while the "inferior" apes are very much alive and well.

FACT (2) Concerning the amount of evidence in the form of bones which supposedly link man and ape, SCIENCE DIGEST said : "The remarkable fact is that all the physical evidence we have for human evolution can still be placed, with room to spare, inside a single coffin".

FACT (3) "Just as we are slowly learning that primitive men are not necessarily savages, so we must learn to realize that the early men of the ice-age were neither brute beasts nor semi-apes, nor cretins. Hence the ineffable stupidity of all attempts to reconstruct Neanderthal or even Peking man".

FACT (4) Modern apes seem to have sprung from nowhere. There is no evidence of any ancestors in the fossil record.

FACT (5) Mans opposable thumb with four fingers gives us great dexterity. We can do all sorts of things with coordination of our brain and hand. How do apes compare ? "The apes, having short thumbs and long fingers are handicapped in relation to delicate manual dexterity." (New Encyl. Brit.)

FACT (6) Brain ability. Researchers at Plymouth university, England, gave six monkeys one computer for a month. The result ? They failed to produce a single word. They produced only five pages of text, primarily filled with a lot of S letters.

Having investigated many different aspects of the theory of evolution in the cold hard light of scientific facts, what conclusion have you reached ? Is the theory worthy of the term MYTH or is it based on truth ?

You be the judge.

CHAPTER 11

In view of the astonishing amount of evidence to discredit evolution, the question remains, if we didn't get here by the process of evolution, how did we get here ?

The purpose of the remainder of this book is take an in-depth look at : Our Universe; the earth; man; animals and plants, and see whether there is any evidence of design. If there is, we will have to conclude their must be a designer. If there is a designer, there must be an intelligent mind. If there is a mind, there must be a body, a person. True this person could not possess a body like ours, as you would expect of someone who could create this amazing universe. This person would need to have a spirit body (non-physical) to create things that would easily kill a human.

If such a person exists, what evidence do we have ? If in our study of the universe we found proof that there is wisdom, power, and even love behind it, would that not support a belief in an intelligent creator, not just some abstract 'thing' but a REAL PERSON.

Just for a moment, consider the house you live in. Look around you.

You see doors, architraves, walls, ceilings, cupboards and tiles etc. You cannot see the plumbing or electrical wires but you know they are there.

You know without any doubt that some person sat down at a desk and drew up plans for the house. That person DESIGNED the size, style, and overall feel of the place. You would never walk in another persons house and assumed it just happened by accident.

You would KNOW it was designed by someone and built by an array of tradespeople.

Would you be prepared to apply this same principle to the universe and the earth in particular ? Only you can answer that. Whatever views you hold, I ask you, for the moment to suspend any scepticism you might have, at least until you have read to the end of this book.

Whatever our beliefs, it would be true to say that when we look up into the night sky and see the many stars and planets with the naked eye, do we not marvel at how awesome our universe is ? And yet what we are able to see is a mere FRACTION of the rest of the universe.

Our solar system is part of a great galaxy of stars called the Milky Way. When we consider the vast distances involved just in our Solar system, we get to understand the vastness of our galaxy. For instance, Pluto is the most distant planet in our Solar system at an average of 5,914,000,000 kilometres from the sun. In round figures, lets just call it six billion kilometres, which at some point it actually is. A spaceship travelling at approximately 25,000 kilometres an hour would take twelve years to get there.

Now try and visualize our Solar system on the edge of our galaxy; so small it just looks like a few dots grouped together. The diameter of our galaxy is so vast, that if you were able to travel at the speed of light (186,282 miles per second) it would take you 100,000 years to cross it. Its almost impossible for our minds to grasp such distances.

Our galaxy is not alone. Some astronomers estimate that there are over ONE HUNDRED BILLION of them scattered throughout the universe, and each galaxy may contain hundreds of billions of stars ! Now you would think that these galaxies are scattered haphazardly in space, but this is not the case. They are actually arranged in definite groups known as clusters. Thousands of these have already been observed and photographed. In the most recent discovery, astronomers have found that galaxies ALL ROTATE IN THE SAME DIRECTION which has left scientists gobsmacked as to why this is.

Our Milky Way galaxy is part of a cluster of about 20 galaxies. One of these, the Andromeda galaxy, can be seen with the naked eye on a clear night. Some galactic clusters are made up of hundreds, even thousands of galaxies. One of them is thought to contain as many as 10,000 galaxies !

The distance between galaxies WITHIN a cluster may average about a MILLION LIGHT YEARS! It doesn't end there because astronomers say there is evidence to suggest that these clusters of galaxies are themselves arranged in "superclusters", a bit like bunches of grapes on a vine. What amazing organization is evident here !

A senior writer for SCIENTIFIC AMERICAN came to this conclusion :

"The more clearly we can see the universe in all its its glorious detail, the more difficult it will be for us to explain with a simple theory how it came to be that way".

Thus the dilemma that faces those who choose to believe the universe 'just happened', without outside influence, why is there so much order and precision, or for want of a better word — DESIGN. Even the great scientist Albert Einstein after years of studying the universe acknowledged there must be a "supreme mathematician" behind it all.

Just think of the mathematics involved of billions of stars all in their own orbit, flying around at amazing speeds and yet not clashing into one another, and this over billions of years ! To work all this out would take a genius without equal.

Astronomers have discovered that our universe is not static as once thought, but is actually expanding. For example, in 1995, scientists were watching the strange behaviour of the most distant star ever observed (SN1995K). This star became very bright and then slowly faded. NEW SCIENTIST magazine plotted this on a graph and explained : "The shape of light curve.......is stretched in time by exactly the amount expected if the galaxy was receding from us at nearly half the speed of light. This is the best evidence yet that the universe really is expanding".

This fact proves that the universe had a beginning and has not always existed as first thought. What this means is that there was a time when there was no matter at all in existence, only energy. This is in line with Einstein's theory that energy can be converted into matter and vice versa.

This raises some important questions : Where did the energy come from ?

Because energy in itself is not very useful unless directed in some way, it leads us to conclude that something must have started the direction of this energy into matter. A force so powerful that it could overcome the immense gravity of the entire universe.

What could be the source of such dynamic energy ? Apart from this question there is evidence to suggest that foresight and intelligence are involved. This is because the expansion rate seems to be very finely tuned as is pointed out by scientist Sir Bernard Lovell : "If the universe had expanded one million millionth faster, then all the material in the universe would have caused the universe to collapse within the first thousand million years or so of its existence. Again there would have been no long lived stars and no life".

Such amazing precision would have to point to an equally amazing intelligence. The implications of this were highlighted by Robert Jastrow, Professor of Astronomy at Columbia University. To quote : "The astronomical proof of a beginning places the scientists in an awkward position, for they believe that every effect has a cause.......The British astronomer E.A.Milne wrote 'we can make no propositions about the state of affairs (in the beginning) in the divine act of creation, God is unobserved and unwitnessed". (33)

To help us understand just how finely tuned the universe really is, we are going to look at four fundamental physical forces that effect the vastness of the cosmos and even the infinitely small atom. They are:

GRAVITY; ELECTROMAGNETISM; STRONG NUCLEAR FORCE and WEAK NUCLEAR FORCE. Without these four forces, elements vital for life would not exist.

Gravity is an incredibly powerful force which can bind stars and planets together and even galaxies. Its power is evident to all of us here on earth, because we are well aware of its effect if for example we were to jump of a cliff. The earths gravitational pull is ideal for life here, as it prevents us from flying off into space. We've all seen films depicting the effect of zero gravity on astronauts. Its quite funny to see them floating around or when on the moon, every step is like a jump. The force of gravity may vary from star to star, but its very finely tuned.

Equally important in the great order of the universe is electromagnetism. If this force were significantly weaker, electrons would not be held around the nucleus of an atom. Ultimately this would mean the absence of chemical reactions between atoms, resulting in the end of all life.

On a greater scale think of the effect that just a slight difference in the electromagnetic force would have on the sun. Because it would alter the light reaching the earth, it could make photosynthesis in plants either difficult or impossible. It could also deny us the unique properties of water, so vital to life.

An additional important aspect of this fine tuning is the intensity of electromagnetism in relation to the other three. For example its been estimated that it is 10/40 times that of gravity. That is ten followed by forty zeros. Thats not just billions of times greater but quadrillions greater. We're talking big numbers here.

What if we added just ONE zero to the above number and made it 10/41 would that seemingly small change make a difference ? Yes it would, because gravity would now be proportionately weaker. What would be the result ?

Scientist Dr. Reinhard Breur tells us : "With lower gravity the stars would be smaller, and the pressure of gravity in their interiors would not drive the temperature high enough for nuclear fusion reactions to get under way : the sun would be unable to shine".

But what if the gravity was stronger in proportion so that there was one less zero and the number of zeros were 39 instead of 40, what would happen then ? Breur continues : "With just this tiny adjustment, a star like the sun would find its life expectancy sharply reduced".

So its clearly evident that there is a lot of fine tuning with regard to electromagnetism. But what about the remaining two forces essential to the running of the universe : Strong and weak nuclear force.

These two forces exist in the nucleus of an atom. STRONG NUCLEAR FORCE has the power to glue protons and neutrons together. Because of this bonding, elements can form. For instance light ones such as helium and oxygen and heavy ones such as gold and lead. What if this binding force was a mere 2% weaker, what would result ? Only hydrogen would exist, meaning an end to all life.

On the other hand if the force was slightly stronger, only heavier elements would exist but no hydrogen. If that happened the sun would not have the fuel it needs to radiate life giving energy, which in turn would mean we would have no food or water. Fine tuning indeed wouldn't you agree ?

The other force mentioned is the WEAK NUCLEAR FORCE. It controls radioactive decay and also effects thermonuclear activity in the sun.

Would it make any difference if it was slightly weaker or stronger than it is now ?

According to physicist Freeman Dyson it would. He says : "The weak force is millions of times weaker than the strong nuclear force. It is just weak enough so that the hydrogen in the sun burns at a slow and steady rate. If the weak force were much stronger or weaker, any forms of life dependent on sunlike stars would again be in difficulties".

Think about it. All the four forces essential to the smooth running of the universe : GRAVITY; ELECTROMAGNETISM; STRONG NUCLEAR FORCE and WEAK NUCLEAR FORCE, are so finely tuned that a very small change in proportions would result in chaos. Does this not point to a brilliant designer come mathematician, rather than blind chance ? You be the judge.

CHAPTER 12

This now brings us to another fascinating subject : Our wonderful planet earth. No place like it has ever been found anywhere in the universe. It truly is deserving of the designation — UNIQUE.

Its position in relation to the rest of our solar system, its measurements, and atmosphere, make it exclusively suited for life to exist.

Take for instance the average distance between the earth and the sun. What if it was a bit closer or further away than it is now ? Would it make a difference ? Professor David L. Block gives the answer : Calculations show that had the earth been situated only 5% closer to the sun, a runaway greenhouse effect would have occurred about 4,000 million years ago. If on the other hand the earth were placed only 1% further from the sun, runaway glaciation would have occurred some 2,000 million years ago".

What about the dimensions of the earth could that alter anything ?

Its thought to be exactly the right size for our existence here. If for example it was slightly larger, its gravity would be stronger, and hydrogen a light gas, would collect, being unable to escape would make life inhospitable. Conversely, if the earth were slightly smaller, life-sustaining oxygen would escape and surface water would evaporate.

Either of the above scenarios happening and we would not be able to live here.

The rotation of the earth around the sun is another example of the amazing precision with regard to our planet. The very fact that the earth rotates on its axis once a day, makes for the moderate temperature we enjoy. As a comparison, look at Venus which takes 243 days to rotate.

Just think if it was like that for us, we couldn't survive the extreme temperatures resulting from such long days and nights.

Even our position in the solar system, which is at the edge of the Milky Way galaxy is important. Were it nearer the center of the galaxy the gravitational effect of neighbouring stars would distort earths orbit.

Another important thing to consider is the speed at which the earth travels through space. It orbits the sun at a speed of 66,000 miles an hour (approx. 126,000 KMS.) This is just the right speed to offset the gravitational pull of the sun, keeping the sun at a safe distance.

If the speed decreased, we would be pulled toward the sun making us a scorched wasteland like Mercury. If our orbital speed increased, earth would move further away from the sun and we would end up a frozen waste like Pluto.

Another interesting fact that makes our planet unique is that as the earth rotates on its axis, it is tilted at 23.5 degrees in relation to the sun.

If this were not so there would be no change in seasons and our climate would be the same all the time. Although we could still survive, it would make life less interesting and in many places cause a drastic change to crop cycles.

Perhaps the most amazing thing that makes our planet unique in our solar system is the atmosphere that surrounds it. No other planet or star has it, which explains why astronauts have to wear special suits when they venture into space.

Our atmosphere has just the right proportion of gases essential for life. But heres the surprising thing : By themselves, some of these gases are deadly; but because the air has them in safe proportions we can breathe without ill effect.

Most of us know what essential gas makes up a good part of the air we take in. Yes, it is of course oxygen, which makes up 21% of what we breathe in. Without it humans would die within minutes. But too much would be harmful, because if breathed too long it becomes toxic. Another problem. If there was too much oxygen in the atmosphere all combustible materials would become highly flammable, creating a nightmare for firefighters.

Fortunately, oxygen is diluted with other gases, especially nitrogen which comprises 78% of our atmosphere. Making up less than 1% is carbon dioxide. It may seem a very small amount but it too is essential for plant life. Without it all plants would die. In a wonderful example of balance, humans and animals breathe in oxygen and exhale carbon dioxide which is taken up by the plants. If that one percent were increased it could be harmful to us and the animals. If it were decreased it could not support life as we know it. What incredible harmony and precision is evident in this cycle of life.

Another function of the atmosphere is it acts a protective shell. About 15 miles (24 km's) above the ground there is a thin layer of ozone gas that filters out harmful radiation from the sun. Evidence of a hole in the ozone layer and the resultant problems down here on earth shows just how important this shell really is. It also shields the earth from bombardment from meteors. Most burn up before they get here, otherwise millions of meteors would strike all over the place, causing untold damage.

To give you an idea of the sort of damage they can do if we lacked this protective atmosphere, consider South Africa. It has the worlds largest crater. Its believed to be made from a meteor which hit earth around two billion years ago, when we assume the earths atmosphere was not in place. It is situated near Johannesburg and believe it or not has an estimated diameter of 186 miles (300 km's) ! Imagine what a meteor like that could do if it hit New York or London !

Besides the protection it gives us, the atmosphere keeps the warmth of the earth from being lost into space. In addition to being vital to life, what wonderful sights we see in an ever changing sky. The golden glow of dawn and the pinks and reds and purples of glorious sunsets. A writer in the NEW ENGLAND JOURNAL OF MEDICINE summed it up when he said :

"Taken all in all, the sky is a miraculous achievement. It works, and for what it is designed to accomplish, it is infallible as anything in nature.

I doubt whether any of us could think of a way to improve it, beyond maybe shifting a local cloud from here to there on occasion". (34)

Yes, as this writer observed, the atmosphere is a miraculous achievement and 'works for what it is DESIGNED to accomplish'.

When you consider the information so far, what do you think ? When you see the precision involved in our earths position in the solar system; the exact distance from the sun; the right speed (126,000 KPH); the exact tilt on its axis (23.5 degrees); the exact amount of oxygen in our atmosphere (21%); just the right amount of nitrogen (78%), do you not wonder at it ?

Would it make sense to believe all this came about by accident ? Or would it be more scientific to believe that behind all this precision is a supreme mind; a very clever, calculating mathematician, who had the ability and wisdom to work it all out, and having the enormous power and energy to create it. Again, you be the judge.

CHAPTER 13

Perhaps the most amazing example of order and precision is to be found in the structure of elements.

I remember when at school in science class, I wondered how they arrived at the numbers alloted to each element. The teacher may have told us, but it didn't sink in. It wasn't until decades later that I discovered the answer. Before we go into this lets consider some facts about them.

About three hundred years ago, only 12 elements were known. They were : ANTIMONY, ARSENIC, BISMUTH, CARBON, COPPER, GOLD, IRON, LEAD, MERCURY, SILVER, SULPHER, and TIN. As time went by and more and more were discovered, scientists noticed that the elements reflected a distinct order. Because of gaps in this order some scientists such as Mendeleyev, Ramsey, Mosely and Bohr, theorized the existence of unknown elements and their characteristics. Eventually those elementswere discovered just as predicted. How could these men so accurately predict forms of elements that were unknown at the time ? Well its because the elements follow a natural numerical order based on the structure of their atoms. This is a proven scientific law. This is why school textbooks can set out a periodic table of the elements in rows and columns in a definite order.

Atoms of elements have, moving around its nucleus, what are calledelectrons, protons and neutrons. The number of protons found (under a powerful microscope) decide where the particular element is placed on the periodic table. Hydrogen for example has only ONE proton, therefore it is the first on the periodic table.

The last element that occurs naturally on earth is number 92, uranium, which has 92 protons.

From the early days of discovery of elements, it became clear to some scientists there were gaps in the table. This led Russian scientist DmitryMendeleyev to predict existence of the element with the atomic number 32, germanium, as well as its colour, weight, density and melting point. Was his prediction correct? The 1995 science textbook CHEMISTRY gives this answer : "Mendeleyev's prediction about other missing elements – galiumand scandium, also turned out to be very accurate".

Other scientists made similar forecasts about missing elements. Eventually, all the missing ones were found. There are no longer any gaps left on the chart.

When we think of the rich variety of chemical elements it is truly fascinating. For example, gold and mercury have distinct shiny colours.

One is a solid, one a liquid. They appear on the chart as numbers 79 and 80. Now isn't it interesting that just one more proton in mercury, can change it from being a solid (as in gold) to a liquid. Evidenceof order and design wouldn't you agree ?

One final consideration of the elements on the periodic table is truly amazing. Scientists discovered a remarkable relationship between elements that share a column. For example, in the last column are locatedthe following elements : HELIUM (No.2); NEON (No.10); ARGON (No.18); KRYPTON (No.36); XENON (No.54); and RADON (No.86). All of these are gases that glow brightly when an electric discharge passes through them, and they are used in some light bulbs. Perhaps the most familiar of these gases is neon, as we have all heard of the neon lights used for commercial advertising. Besides this common factor uniting these elements, they all do not react easily with various elements as do other gases.

Whether the preceding examples of order and precision were accidental or the result of a deliberate act, it would be appropriate to look at the views of some knowledgeable researchers. For instance, in 1988, a book that attempts to explain how life could have arisen by chance was reviewed in the journal SEARCH by the AUSTRALIAN & NEW ZEALAND ASSOC. FOR THE

ADVANCEMENT OF SCIENCE. On just ONE page of the book, science writer L.A. Bennet found "sixteen highly speculative statements, each depending on the preceding for credence". After reading the whole book, what was his conclusion ?

"It is far easier to accept an all-loving creator, instantaneously creating life and guiding it along its teleological pathways....... than to accept the myriad 'blind chances' needed to support the author's theses".

Fine-tuning is evident everywhere we look, which caused mathematician and astronomer David Block to draw a similar conclusion when he said : "We live in a very finely tuned universe....... our universe is a home. Designed I believe by the hand of God".

CHAPTER 14

Imagine the perfect factory. It would be non-polluting, quiet, pleasant to the eye, super efficient, inexpensive to run and not dependent on coal or oil to power it. This factory has the most advanced technology imaginable without all the unexpected glitches, breakdowns or endless tweaking that cutting edge technology requires.

Impossible you say ? No. Every minute of every day this incredible factory is at work and it goes by the name of photosynthesis. Its been estimated that altogether, earths green plants produce an approximate amount of 150 to 400 BILLION TONS OF SUGAR EVERY YEAR !! To give you an idea of how great this production is, the aforementioned figure is far more material than the combined output of all this worlds iron, steel, automobile and aerospace factories.

How do plants manufacture this amazing output ? They do this by using the energy from the sun to remove hydrogen atoms from water molecules and then attach them to carbon dioxide molecules from the air, turning it into a carbohydrate known as sugar. The plants can now use their new sugar molecules for energy or combining them into starch for food storage, or use them for making cellulose (the tough stringy material that makes up plant fibers).

Next time you look up at an enormous tree such as a sequoia, remember that apart from water and a few minerals, it was largely made up of thin air. Sounds like magic, but in reality is evidence of intelligent design. VERY INTELLIGENT. Scientists are only now starting to understand just how sophisticated the process really is.

When researching this process, I found it just too complex to explain in simple terms and therefore admit defeat. Without going into detail, I will just touch on one aspect of photosynthesis and that is : WHY THE GRASS IS GREEN.

As sunlight strikes a plant a microscopic thing called a thylakoid containing chlorophyll molecules, is waiting to snare it. These molecules are especially interested in absorbing red light of a specific wavelength. Meanwhile other molecules are absorbing blue and violet light.

Now this is where it gets really interesting. Of all the wavelengths falling on plants, only green light is useless to them. In truth green light is a waste product of plants, and yet what joy this brings to the eyes, especially in spring and summer.

AS mentioned earlier, the full process of photosynthesis is just too complex to explain in understandable language. But this very fact in itself proves there has to be intelligent design behind it. One thing I forgot to mention is a by-product of the process which is vital to our existence :

OXYGEN. So important is this that the NEW ENCYLOPAEDIA BRITANNICA had this to say : "Without photosynthesis not only would replenishment of the fundamental food supply halt, but the earth would eventually become devoid of oxygen".

Although scientists have attempted to explain the step by step process in their textbooks, some steps, even now are not fully understood. Evolutionists cannot explain how each step evolved from something simpler because each step appears irreducibly complex.

Some world famous scientists find it hard to believe that photosynthesis is the result of blind chance. At a meeting of the AMERICAN PHYSICAL SOCIETY, Nobel prize winner Robert A. Millikan had this to say : "Theres a divinity that shapes our ends......a purely materialistic philosophy is to me the height of unintelligence. Wise men in all ages have always seen enough to at least make them reverent".

In his speech, to prove his point he quoted the great scientist Albert Einstein who confessed : "I try humbly to comprehend even an infinitesimal part of the intelligence manifest in nature".

Yes, intelligence indeed. Another remarkable action in the plant world is their circulatory system. We've discussed the benefits of photosynthesis but it cannot function without the help of tiny roots. Each root tip is fitted with a protective cap, each lubricated with oil, can then push their way up the soil. Root hairs behind this oily cap absorb water and minerals, which travel up minute channels in the sapwood to the leaves. Once there sugars and amino acids are made.

This circulatory system is so amazing that many scientists regard it as virtually a miracle. For example, just think about it, water and nutrients are sometimes pumped up to a hundred metres above ground as is the case with trees like the sequoia. How does it do this ?

First of all, root pressure starts it on its way but when it gets to the trunk another mechanism takes over. Water molecules hold together by cohesion, as water evaporates from the leaves (known as transpiration) the tiny columns of water are pulled up like ropes, travelling up to 70 metres an hour. Its been said that this system, believe it or not, could lift water up to two miles high (3.5 kms)!

As excess water evaporates from the leaves, billions of tons of water are recycled into the air, only to fall later as rain. Does not this wonderful system give evidence of design ? You be the judge.

Finally, in our consideration of plants, lets take a look at its very genesis......the seed. They come in a bewildering variety of shapes and sizes. Orchid seeds are so light they float off like dust. Dandelion seeds come equipped with parachutes. Maple seeds have wings and flutter off like butterflies. Some plants have pods that snap open and the seeds are catapulted out.

Some desert plants have seeds that only sprout if at least half an inch of rain has fallen. The seed even seems to know which direction the rain is coming from. If water comes from above they will sprout, if from below it won't. There are harmful salts in the earth that need to be leached out if the seeds are to sprout, hence the seed will only germinate if water comes from above.

Amazingly the biggest living thing on earth, the giant sequoia tree, grows from one of the smallest seeds.

It can grow to over a hundred metres in height. Just over a metre above the ground its diameter may be as much as 12 metres (36 ft.). There may be enough timber in one tree to build fifty six room houses! Its roots cover 3 to 4 acres, and its known to live for over 3,000 years. Remember all this comes from a seed not much bigger than a pinhead. Does that not strike you as evidence of design ?

CHAPTER 15

Bionics is the study of living things, and copying them. Yes, inventors often only repeat what animals and plants have been doing for milleniums. As one biologist put it : "I have the suspicion that we're not the innovators we think we are; we're merely the repeaters". (35)

On this same point another scientist made the comment : "In many areas human technology is still lagging behind nature". (36)

Long before man installed air-conditioning in their homes, termites cooled theirs as they still do. The nest is in the centre of the mound and underneath it is an air chamber which allows fresh air in where it circulates around the actual nest where the termites live. In hot weather water brought up from below evaporates cooling the air. Stale air diffuses out the porous sides of the nest. The question is : How can millions of blind workers co-ordinate their efforts to build these amazing structures ?

Biologist Lewis Thomas gives this answer : "The plain fact that they exhibit something like collective intelligence is a mystery". (37)

Is it not possible that this 'intelligence' was programmed into them from the very beginning by a superior intellect ?

We humans were not the first to use antifreeze, like the glycol we use in our car radiators. Microscopic plants that inhabit Antarctic lakes use chemically similar glycerol to prevent themselves from freezing. This chemical is also found in insects that can survive temperatures of four degrees below zero fahrenheit. Fish too that live in the waters around the Antarctic also are endowed with their own antifreeze.

Humans that sometimes go diving, use air tanks strapped to their backs so they can breathe under water for up to an hour, but we are not the first to do this. Water beetles use a simpler method and can stay down longer. They grab a bubble of air and submerge. The bubble acts like a lung. It takes carbon dioxide from the beetle and diffuses it in the water and then takes oxygen dissolved in the water for the beetle to use.

Long before humans had clocks or sundials, living organisms were keeping accurate time. For example, when microscopic plants called diatoms come to the surface of wet beach sand when the tide is out. Guess what ? If you put these tiny creatures in a laboratory in a pile of sand without any tidal ebb and flow, their internal clocks will still make them come up and down in time with the tides! A similar thing happens with fiddler crabs.

What about birds ? They can navigate by sun and stars which change position as time passes. To compensate for this, they would have to have some sort of internal clock.

Because water is a valuable commodity, there has been much talk in Australia of building a huge desalination plant in Sydney. To accomplish the process of desalination, massive factories have to be built. It is an extremely complex affair. Were we the first to come up with this idea ?

Apparently not. Mangrove trees that live right on the edge of seawater have roots that suck it up, then filter it through membranes that remove the salt.

Birds such as seagulls, pelicans, cormorants and the albatross, all drink seawater and remove the excess salt by means of special glands in their heads.

What about electricity, were we the first to harness it ? For the answer we need to look no further than beneath the ocean. Believe it or not, but some 500 varieties of electric fish have built-in batteries. The African catfish can produce 350 volts ! Shocks from the South American electric eel have been measured as high as 886 volts ! !

Alright you say, so they had electricity first, but what about jet propulsion, surely we were first in that field. Alas no. The octopus and the squid both had it first, milleniums before Frank Whittle invented the jet engine. With aircraft, the jet engine sucks air in the front and blasts it out the back, causing forward thrust. Works a treat. As good as it is, we have to concede that the octopus and squid are the masters of this form of motion. They suck water into a special chamber and then with powerful muscles, expel it, shooting themselves forward. These two creatures don't have the monopoly on jet propulsion, as others with this ability are as follows : The chambered nautilus; scallops; jellyfish; and dragonfly larvae.

Thousands of years ago the Egyptians made paper but they were not the first. Wasps and hornets had been making it milleniums earlier. To make their nests they chew up weathered wood which produces grey paper which when used in several layers is very tough. Because each layer is separated by dead air spaces, this insulates the nest from heat and cold, as effectively as a brick wall 16 inches thick (406mm).

When the rotary engine was first brought to the publics attention, there was much hoo-ha surrounding it. Undoubtedly it was a good invention and a big step forward in engineering, but it wasn't the first. It was preceded by microscopic bacteria aeons earlier. One type of bacteria has hairlike extensions twisted together into a stiff spiral, a bit like a corkscrew. It spins this around like the propeller of a ship and drives itself forward. And how about this —— it can even reverse its engine ! How all this works is not completely understood and may never be.

The speed this bacterium can travel at is estimated to be the equivalent of 60 kilometres an hour. As one researcher marvelled : "One of the most fantastic concepts in biology has come true : Nature has indeed produced a rotary engine, complete with coupling, rotating axle, bearings, and rotating power transmission". (38)

Another glaring example of who had it first is that of sonar. Not only did bats and dolphins have it long before man, their version is far superior.

Take bats for instance. In experiments where scientists set up a darkened room with fine wires rigged across it, bats fly about and never touch the wires. The bats give out supersonic sound signals that bounce off these objects and instantly return to them, making use of echolocation to avoid the obstacles. Porpoises and whales do the same thing only in water.

History might have turned out differently if British scientists hadn't made this discovery before the Germans in the second world war. It was their early use of radar (learned from bats) that enabled spitfires to be in the air before the Luftwaffe arrived. If the RAF had lost the battle of Britain due to the absence of an early warning system, Hitler would have launched his "Sealion" invasion, probably won, and America would have had no base in which to launch an invasion of Europe. So, perhaps its fair to say that the humble little bat had a part in the outcome of the war.

From the 17th century onwards, men have been developing thermometers. Yet, as good as they are, they are crude in comparison to those found in nature. The beaks of a Mallee bird and a brush turkey can tell temperature to within one degree fahrenheit (–17 deg. Celsius).

The antennae of a mosquito can sense a change of 1/300 of a degree fahr. Amazing as this is, the king of all natural thermometers has to be the rattlesnake. At the side of its head it has small pits which can sense a change in temperature of just 1/600 of a degree fahrenheit !

At this point, its advantageous to examine the acute senses inherent in birds, insects, fish and animals. For instance consider the sight of birds and insects. We humans cannot see ultra violet light, but many birds and insects can. Bees for example can orient themselves in relation to the sun even on a cloudy day, by locating some blue sky and than checking out the pattern formed by polarized U V light. Even flowering plants make use of patterns that are only visible in the U V range which acts as a 'nectar marker' to point insects in the right direction. Certain fruits and seeds use this form of advertising to attract the attention of birds.

U V light is believed to be an aid in helping certain hawks and kestrels to locate voles and field mice. According to the journal BIOSCIENCE, "male voles produce urine and faeces containing chemicals that absorb UV and mark their trails with urine". This makes it easy for birds of prey to hone in on areas of high vole density and concentrate their efforts there. How kind of the voles to give their enemies a sporting chance.

When it comes to eyesight, the vision of birds borders on the miraculous. Why is this so ? The answer is found in a book called ALL THE BIRDS OF THE BIBLE : "The chief reason is that the image forming tissue lining the eyes interior is richer in visual cells than eye of other creatures. The number of visual cells determines the ability of the eye to see small objects at a distance. While the retina of a humans eye contains some 200,000 visual cells per square millimeter in comparison birds can have up to a million receptors in their eyes". Added to this, some birds have the extra asset of two fovae (areas of maximum optical resolution) per eye, giving them a superior perception of distance and speed. Birds have an unusually soft lens that enables rapid focus. Just imagine how dangerous it would be flying through a forest or thicket if everything were a blur. Rapid focus is essential for birds in flight to avoid potential hazards.

Really, would it not make sense to believe that the avian eye is the product of thoughtful design rather than blind chance ? For an eye to be of any use to either bird or man it had to be complete at the outset. A partial eye would be a serious disadvantage, especially to a bird.

Even Darwin himself admitted : "To suppose that the eye with all its inimitable contrivances for adjusting the focus to different amounts of light and the correction of spherical and chromatic aberration, could have been formed by natural selection, seems , I freely confess, absurd in the highest degree".

Yes, absurd indeed. But in spite of this admission, Darwin tried to explain that evolution produced the eye by many transitional stages. Really though does it make sense ? Think for a moment of any bird or animal having an eye that was only a quarter or half formed. With hazy vision, how on earth is the creature going to survive in order to pass on any improvement to its offspring. Even birds with 20/20 vision, sometimes battle for survival, so what hope would any creature have which was half blind ? Its completely nonsensical to think that the eye evolved it simply is impossible.

Still on birds, have you ever wondered how they are able to balance so well when walking along a branch or the narrow top of a fence or even a telegraph wire ? Because their bodies are specially oriented horizontally to enable them to fly, how do they do this balancing act ? "Their tails are not an adequate counterweight" says the LEIPZIGER VOLKZETUNG from Germany. "After four years of research, animal physiologist Reinhold Necker succeeded in finding a second organ of equilibrium in pigeons".

The paper went on to explain that Necker discovered nerve cells and cavities containing fluid in the pelvic region of birds which evidently control the balance. The report continued : "When the fluid spaces were opened, the pigeons were no longer able to sit erect or walk once their eyes were covered. They fell from their perches or toppled over on their sides. Yet they were still able to fly". This was because a separate organ of equilibrium in the inner ear which coordinates their movements in flight was unaffected.

Isn't this simple fact proof of design ? You be the judge.

Hearing is another area where animals have it over us. Most dog owners will tell you that their dog can hear a pin drop at twenty paces. Whereas we can hear sounds ranging from 20 to 20,000 herz, (cycles per second), dogs can hear in the range of 40 to 46,000 herz, and horses between 31 to 40,000 herz. Elephants and cattle can even hear in the infrasonic range (just below human hearing) to as low as 16 herz. The advantage of this is that low frequencies travel further, which explains how elephants can communicate over distances of 3 km or more. Some researchers have suggested that we use these creatures as an early warning of earthquakes and severe weather disturbances, both of which emit infrasonic sound.

Insects are endowed with hearing in both the infrasonic and ultrasonic range. Some have eardrum-like membranes all over their body except their head. Others have delicate hairs that respond to both sound and the most gentle movements in the air, such as those caused by the human hand. No wonder we find it so difficult to swat a fly !

Perhaps the most acute hearing belongs to the bat. Amazing as it may seem, it can actually hear an insects footstep ! We know that bats navigate in the dark and catch insects by means of echolocation or sonar.

Describing how the bat uses this method I quote Professor of psychology H.C. Hughes : "Imagine a sonar system more sophisticated than that found in our most advanced submarines. Now imagine that system is used by a small bat that easily fits in the palm of your hand. All the computations that permit the bat to identify distance, the speed, and even the particular species of insect target are performed by a brain that is smaller than your thumbnail!"

The precision of echolocation depends on the quality of the sound emitted. According to one reference work, bats have "the ability to control the pitch of their voice in ways that would be the envy of any opera singer". To cap it all, these unique little creatures are equipped with a highly sophisticated system for detecting its own echoes. This becomes necessary at night when thousands exit a cave to hunt insects.

If they were unable to detect their own echo, confusion would reign and none would get fed. Evidence of clever design don't you think ?

As clever as man is at navigation, we can still learn something from the natural world. "Spiny lobsters have an uncanny ability to find their way home even after being blindfolded, driven in circles and plunked down in unfamiliar waters," states Canada's NATIONAL POST. Researchers captured dozens of lobsters off the Florida Keys, placed them in dark tanks and released them up to 23 miles from where they were caught. Though their eyes were covered, the lobsters always gravitated toward their place of capture. The researchers suggest that this is the most advanced form of navigation yet found in an invertebrate.

Head of research Dr. Kenneth Lohmann said : "no matter what we did, the lobsters figured out the direction they needed to walk in order to go back home. Its really a rather remarkable finding if you think about it........

these little crustaceans being able to somehow determine their position under conditions in which humans would be completely lost".

CHAPTER 16

This chapter is dealing with a strange and marvellous thing — INSTINCT. In the world of nature, when birds and other creatures act purely from instinct, are we not astounded and even perplexed as to the origin of these mental powers ? How do birds know WHEN to migrate to another area ?

How can they unerringly end up at the same location as they were the previous year ? How do birds know how to build a nest when it was never taught them by their parents ? The answer of course is instinct. Perhaps this is the one area that sets humans and animals apart. Basically we have to be taught how to eat, walk, communicate and orientate ourselves. It does not come naturally.

This instinctive wisdom inherent in birds, animals, fish and insects, created an enormous dilemma for Darwin. He wrote : "Many instincts are so wonderful that their development will probably appear to the reader a difficulty sufficient to overthrow my whole theory......I may here premise that I have done nothing to do with the origin of the mental powers, any more than I have with life itself". (39)

Yes, Darwin admitted he couldn't explain the origin of instinct anymore than he could the origin of life. They were both a mystery. Over a century later, are scientists any closer to understanding how this instinctive behaviour arose in animals and birds ? Notice this comment from evolutionist Gordon Rattray Taylor : "The plain fact is that the genetic mechanism shows not the slightest sign of being able to convey specific behaviour patterns......When we ask ourselves how any instinctive pattern of behaviour arose in the first place AND BECAME HEREDITARILY FIXED we are given no answer".   (40)

What this man is saying is that although we understand how certain things can be passed on genetically, (things like eye colour, feathers, scales or hearing quality etc.) how does an abstract thing such as behaviour patterns get passed on and fixed permanently. Its true that certain behavioural traits in us humans can get passed on genetically, inherited from parents or even grandparents, but this does not become a PERMANENT fixture as is the case in the natural world.

Some evolutionists try to explain the phenomenon of instinct by saying that with bird migration for example, to find a better climate and in search of food, the birds wandered experimentally over large areas.   But scientists are well aware that learned behaviour is not incorporated into the genetic code and hence is not inherited by the offspring. They are forced to admit that migration is instinctive and "independent of past experience". To prove this point lets consider a few examples.

When it comes to long distance travel, the Arctic Tern is the champion. During summer it nests way up north of the Arctic Circle. At summers end they fly south to spend the Antarctic summer near the South Pole. They may circle the entire continent (a huge area) before heading north to the Arctic. The annual round trip of this migration is around 33,000 kilometers ! They do this trip because of the rich sources of food at these two extremes during the summer months. One scientist posed the question : "How did they ever discover that such sources existed so far apart"? Evolution has no answer.

Just as hard to explain is the migration of the Blackpoll Warbler. It is very light weighing only three quarters of an ounce. In autumn it travels from Alaska to the eastern coast of Canada or New England; gorges on food, stores up fat and then waits for a cold front. On its arrival, the bird takes off to its final destination being South America. But this is where it gets really interesting. Instead of flying in a straight line, it first heads toward Africa. On the way, somewhere out over the Atlantic ocean it picks up a prevailing wind that turns it toward South America. To pick up this wind they need to fly at an altitude of around 30,000 meters. This raises a number of pertinent questions : How does the Warbler know to wait for the cold front that means good weather and a tail wind ? How does it know to keep climbing ever higher where air is thin and has 50% less oxygen ? How does it know its only at that height that the cross-wind blows that will carry them to South America ? How does it know to fly towards Africa as along the way they pick up the southwestern drift from this wind ?

On this trip of 3,600 kilometers, the Blackpoll does not consciously know any of the things it does. It does them purely from instinct. But where does this instinctive behaviour come from ? It could not possibly have evolved as its not learned either from experience or from parents. It had to be there from the very beginning of their existence, which would make sense if some supreme being of incredible intelligence and wisdom put it there at the very moment He made them. This is the only reasonable explanation for instinct in animals and birds, which must have been programmed into them at the beginning.

As proof of this, lets look at two examples. First the New Zealand long-tailed cuckoo. The young travel from their home a distance of around 6,000 kilometers to certain Pacific islands, where they join their parents who had gone earlier. Now think about this : These young birds had never been to this destination before and yet were able to fly to the exact location of their parents, over featureless seas and over a vast distance. What further proof do we need that this instinctive behaviour must have been programmed BEFORE BIRTH.

A second example is the Manx Sheerwater that annually migrate from Wales to Brazil, leaving behind their chicks, which follow them as soon as they can fly. One of them made the trip in sixteen days, an average of 690 kms per day.

An experiment was once made to see just how good their homing instincts really are. One was taken from Wales to Boston in America, which is far off its migratory route. Yet it returned to its home in 12 and a half days, a distance of 4,800 kms.

Now the point is, if the aforementioned birds can reach their destination without getting lost it can only mean one thing : They have a built-in map inside their heads. Not only that, but to perform these amazing feats of navigation they must use the location of sun and stars to help find their direction. Experiments conducted, seem to confirm this is the way they navigate. And because stars and the sun are not always in the same position, it appears that birds have an inbuilt clock to compensate for the movement of these heavenly bodies.

But lets return to the most fascinating thing of all : The "map" inside their heads. It not only needs the destination on it but also the starting point and the route carefully marked. Of course none of this helps unless they know WHERE THEY ARE LOCATED ON THE MAP. Hence, the Manx Sheerwater taken to Boston, had to identify where he was before he could determine the direction of Wales !

Are you starting to get the picture of how amazing and wonderful is this thing called instinct ? Again I repeat my argument : Such a marvellous thing had to be there from the word go and realistically had to be the product of an intelligent mind. In a sense you could say that in view of the above, GPS (Global Positioning System) is nothing new. The birds had it first.

Six centuries before our common era, the Bible spoke of bird migration long before it was accepted as fact. To quote the writers words:

"The Stork in the sky knows the time to migrate, the dove and the swift and the wryneck know the season of return".   (Jeremiah 8:7)

While still on the subject of birds, consider some other instincts peculiar to them. Take geese for example. We've all observed them flying in a vee formation. Ever wonder why ? There are two basic reasons according to the experts. One is, they fly in formation to keep one another in view so as to respond quickly to changes in direction by the leading bird.

The other reason is believed that the air current created by the geese up front makes flight easier for the rest of the flock, reducing air turbulence.

The adults take turns in leading, as sometimes you can observe the leading bird fall back to the rear so all can get a spell from the harder work at the front. This behaviour is not learned, its instinctive and was there before birth.

The mystery of instinct is not confined to migration and flight patterns.

Nest building is another stunning example of instinct at work.

Science writer G.R.Taylor wrote concerning the genetic machinery involved in this and concluded : "There is not the faintest indication that it can hand on a behavioural programme of a specific kind, such as the sequence of actions involved in nest building". (41)

Its true it cannot be handed down by teaching, but there is no denying its passed down genetically as we we can see by the following examples:

——HORNBILLS of Africa and Asia——

The female finds a cavity in a hollow tree then brings clay, and walls up the opening until she can barely squeeze through. The male brings more mud and he closes the hole until only a slit remains open. Through it the male feeds her and the babies when hatched. When the male can no longer bring enough food, the female breaks out. And guess what, THE BABIES REPAIR THE OPENING. From then on both parents bring food to them.

After several weeks, the babies break down the walls with their beaks and leave the nest.

——SWIFTS——

One species makes its nest out of saliva. The salivary glands of the bird swell and produce a viscous mucous secretion just before the breeding season begins. With its arrival comes the instinctive wisdom of what to do with it. They smear it onto a rock face and as it hardens, adds more layers, and finally a cup shaped nest is the result....How does it KNOW to put the saliva to this use ?

——THE TAILORBIRD——

This bird of Southern Asia is aptly named. It makes thread from cotton or bark fibers and spiderweb, splicing short pieces together to make longer lengths. With its beak it punches holes along the two edges of a large leaf. Then using its beak as a needle, with the thread it pulls the two edges of the leaf together, the way we lace up our shoes. When it comes to the end of the thread it either knots it to hold fast or it splices on a new piece and continues sewing. Hence the Tailorbird turns the big leaf into a cup which then becomes the nest. IS NOT THIS AMAZING ?

——THE HORNED COOT——

This bird usually builds a small flat island. Where it lives however, this type of island is very rare so it makes up its own island, (I'm not kidding.) First it picks out an appropriate place on the water and then begins to carry stones there in its beak. These are piled up in water which is about 1,500 cm to a meter in depth until an island is formed. Amazingly, the base may be as much as four meters in diameter and the pile of stones may weigh more than a tonne. On this stone island the Horned Coot brings in vegetation to build the nest.

The above methods employed for nest building are peculiar to each type of bird. The Horned Coot will always use his island building method as he has for milleniums. The same goes for the Tailorbird. He's been sewing leaves together since antiquity and will continue to do so in the future.

Parents NEVER teach their young their methods of nest-building, the young birds just KNOW how to do it from instinct.

This behaviour could not possibly have evolved, because LEARNED behaviour cannot be genetically passed on. The distinctive habits of birds and animals that are carried on instinctively, must have been programmed into them at the very genesis of their specie.

The maps in their heads, the ability to read them, the best routes to travel, and the most practical nests to build, are all the result of deliberate design. There is no other answer, it had to be this way. This instinctive wisdom could only come from a superior intellect or for want of a better word —— a genius. A brain that knew not only how to embed this wisdom into the makeup of these creatures, this thing we call INSTINCT, but also how they could pass this on to future generations. It may forever remain a mystery as to how this is done. It reminds us of the words of wise King Solomon when he said : "He has given me a sense of time past and future, but no comprehension of God's work from beginning to end".

CHAPTER 17

This chapter is devoted to the human body, and certain aspects of it I am sure will leave your mouth open in awe. Some of our body parts are simply so amazing in their function that its hard to get your mind around it. Take for example our most wonderfully complex organ —— the brain.

Every SECOND some hundred million bits of information pour into the brain from the various senses. Now you would think it would be hopelessly buried under this avalanche of information. How on earth does it cope ?

The answer is truly one of the many wonders of the human brain.

Two factors are involved. First, in the brain stem is a network of nerves the size of your little finger. It acts as a kind of traffic control center, monitoring millions of messages coming into the brain, sifting out the trivial from the essential. Each second this little network of nerves permits a few hundred at most to enter the conscious mind.

The second factor, a further pinpointing of our attention, appears to come about by waves that sweep the brain 6 to 12 times a second. Because of high sensitivity created by the waves, the brain notes the stronger signals and acts on them. Its believed that by this means the brain scans itself, focusing on essentials. Of course we are never conscious of all this flurry of activity that goes on EVERY SECOND.

Just how staggeringly complex the brain is, is shown by this description by French neurobiologist Dr. Jean-Pierre Changeux :

"The human brain makes you think of a gigantic assembly of tens of billions of interlacing neuronal spiders webs, in which myriads of electrical impulses flash by, relayed from time to time by a rich array of chemical signals.

The anatomical and chemical organization of this machine is fantastically complicated". (42)

We tend to take for granted this organ that we use every day in too many ways to mention. Its not that big in comparison to the rest of the human body, weighing on average 1.4 kilos (3pounds). In appearance it resembles a shelled walnut and is protected by our cranium. It is made up of 100 billion neurons or nerve cells, and each one of these may have over a thousand points of contact (called synapses).

Dr. Changeux calculates there are "something like 600 million synapses per cubic millimeter". In other words they would fit onto a small pinhead. How many could be in the average brain ? According to neurobiologist Dr. Richard Restak, "There may be from ten trillion to one hundred trillion synapses in the brain, and each one operates as a tiny calculator that tallies signals arriving as electrical impulses". (43)

Eat your heart out Bill Gates !

But this is not the end of it though. The brains neurons make connections by means of branching fibers called dendrites. Concerning these Dr. Restak makes this observation : "Estimates of the total length of dendrites within the human brain exceed several hundred thousand miles".

The question remains : How is information transmitted within the fascinating universe of our brain ? It is by the conversion of an electrical impulse into a chemical signal that bridges the gap as a neurotransmitter. There are dozens of different chemicals that carry out this job.

All this is sometimes referred to as the brains 'wiring'. The vast numbers of microscopic nerve fibers are precisely placed within a maze of staggering complexity. But how are they placed in the exact spots necessary to fit the 'wiring diagrams' still remains a mystery. This can be seen from a comment by one scientist when he said : "Undoubtedly the most important unresolved issue in the development of the brain, is the question of how neurons make specific patterns of connections......Most of the connections seem to be precisely established at an early stage of development". (44)

Another researcher speaking about these mapped out areas of the brain, said they "are common throughout the nervous system, and how this precise wiring is laid down remains one of the great unsolved problems". (45)

As if this was not complicated enough, scientists have discovered microcircuits that are set up directly between the dendrites themselves. Says one neurologist "These microcircuits are a totally new dimension to our already mind-boggling conception of how the brain works". (46)

An example of how 'mind boggling' the brain can be is its capacity to retain information. How much do you think it could contain ? Twenty volumes of Encyclopedia Britannica ? Carl Sagan, an authority on the brain, states that the brain is capable of holding enough information to "fill some twenty MILLION volumes, as many as the worlds largest libraries". (47)

When you compare the brain with a computer, what P.C. Could possibly contain this amount of information. No competition here.

Of all the organs in our body, the brain sets us apart from the animals more than anything else. To quote one scientist : "What distinguishes the human brain, is the variety of more specialized activities it is capable of learning". (48)

In humans there are many built-in capacities for learning but not the learning itself. In contrast, animals have 'hardwired' instinctive wisdom, but a very limited capacity to learn new things. According to the book THE UNIVERSE WITHIN, the most intelligent animal "never develops a mind like that of a human being. For it lacks what we have : Pre- programming of our neural equipment that enables us to form concepts out of what we see, language out of what we hear, and thoughts out of our experiences". (49)

Even though the ability to learn is pre-programmed, our personal input is necessary to expand our intellect. If we never learned from experience or built on previous knowledge, hardly a trace of intellect could be found.

So our ability to construct our own and program it as we choose, will decide the level of our intelligence as we grow older.

Take for instance a great brain such as that belonging to Albert Einstein. After he was born, he would have made those funny sounds peculiar to all babies and would have had very limited intelligence. His parents would have had to teach him to walk and talk and all those other things infants must learn. If as he grew older and went to school, and had no aptitude for learning and like most children 'hated maths', he could well have become a butcher or a baker. Now I'm in no way demeaning the intelligence of butcher's and baker's, but use them to illustrate that their particular trade would require far less knowledge of mathematics than say a scientist.

Somewhere along the line, Einstein must have discovered a liking for things mathematical. Having found this he would have kept building on this knowledge until eventually he came up with his theory of relativity. Here is a classic case of a person choosing to program his own intellect.

Contrary to the opinion of some, the brain of a black man or an oriental, is in no way inferior to a white man. If given the opportunity of learning, no matter what a person's origin in life, he can develop a great intellect if he so chooses.

As an example lets look at the case of a negro scientist who lived in the 19th century who you've probably never heard of..... George Washington Carver. He nearly died at birth and as a result was a sickly child. Living in the Southern States of America, where it was usual back then to send their children to work as early as possible, Carver's parents faced a dilemma as to what to do with the boy. They soon discovered he loved books and had an aptitude for learning. By much sacrifice from his family he was enrolled in an agricultural college.

Eventually he developed over a hundred uses for the peanut, which single-handedly established the huge peanut farming industry. Because of the peanut farmers success, the cotton industry went into decline. With his constantly improving knowledge and intellect, Carver was able to help the cotton farmers improve their yield by using better techniques that he came up with.

If it were not for this mans dedication to learning, and his willingness to share his knowledge with others, there would be a number of things missing in our pantry, not the least of these are peanut butter and peanut oil. I give Carver's example to make the point that no matter who we are and how we started out on the road of life, unlike animals we can all develop our intellect to whatever level we choose. In this all men are equal.

One of the most fascinating thing about our brain is our memory. Just think, you could be walking along a country road perhaps the first time in many years, and suddenly your nostrils twitch and you recognize a perfume that has not passed your nose since childhood. Your brain does a quick permutation, searching your memory bank to identify the origin of the scent. And lo and behold it tells you it belongs to a particular kind of wild rose. Isn't it amazing that the memory is capable of storing up different scents and is able to identify them after many years without exposure to that particular smell.

Interested in finding just what the the human memory is capable of, I turned to the GUINNESS BOOK OF RECORDS. According to this fine reference work, the famous historian Thomas Macaulay (1800-1859) could recite from memory all of Milton's ten book epic poem Paradise Lost by the time he was fifteen. When it comes to memorizing music, Mozart was without peer. After hearing the Miserere by Allegri only once, he was able to go home and write the complete score from memory !

Dave Farrow of the United States, memorized on a single sighting, a random sequence of 52 separate packs of cards, a total of 2,704, all of which had been shuffled together. He had only six errors at his attempt at the record in 1996.

Although its true, as we get older we complain "My memory is not what it used to be" and its a fact it can fade with age but not always. The famous South African politician Jan Christian Smuts (1870-1950) memorized 5000 books in his old age, even though he did not learn to read until he was 12 years old.

When it comes to mental calculation, German Johann Dase (1824-61) was the king. He was by far the fastest of all mental calculators. His feats included being able to calculate pi to 205 places in his head, multiplying two 8 digit numbers in 54 seconds and two 20 digit numbers in six minutes !

Yes, the human brain is an incredible organ. To say that it evolved from a lower animal goes against all the facts. A far more logical conclusion was reached by Neurosurgeon Dr. Robert J. White when he said "I am left with no choice but to acknowledge the existence of a superior intellect, responsible for the design and development of the incredible brain-mind relationship, something far beyond mans capacity to understand.......I have to believe all this had an intelligent beginning, that someone made it happen". (50)

CHAPTER 18

Let us now consider another wonderful organ, our heart, and see whether there is any evidence of design in it. The heart is primarily a pump but what an awesome pump it is, faithfully gushing out the red stream of life every second of our lives. It weighs less than half a kilo (1.1 pounds) yet it can beat up to a hundred thousand times a day, pumping blood through the body's cardiovascular system (over 90,000 kms in length) to the tune of 7,600 liters a day (2000 gallons). During a lifetime this turns out to be tens of millions of liters pumped throughout the body. And think, some people live up to 127 years, with that little organ pumping away day in day out.

No muscle in the body works harder, longer and steadier, decade after decade than the heart. Its beat is initiated by a concentration of cells making up its pacemaker, sending out electrical impulses that govern the rate of the heartbeat. Only the brain needs more nourishment than the heart.

At this juncture, its appropriate to consider the qualities and functions of our blood which the heart relentlessly pumps around the body. It is made up of both red and white cells each having a specific purpose. The red cell is the most common of the two. Just one drop of blood can contain hundreds of millions of them. When seen under a powerful microscope, they look like doughnuts with a depressed center instead of a hole. Packed into each cell are hundreds of millions of hemoglobin molecules. In turn each of these molecules are made up of about 10,000 hydrogen, carbon, nitrogen, oxygen and sulphur atoms, plus four of iron which gives it its oxygen-carrying capacity. The hemoglobin helps remove carbon dioxide from the tissues, transporting it to the lungs where it is exhaled.

The outer covering of a red blood cell, called a membrane, has a vital part to play in the bloods circulation. Because of its special elasticity, it enables the cells to stretch into thin shapes so as to pass through your tiniest blood vessels, sustaining every part of your body.

Unlike other cells, red blood cells have no nucleus. The advantage of this is it gives them more space to carry oxygen and makes them lighter which helps your heart pump all those trillions of cells through the body.

Because of lacking a nucleus they are able to renew their internal parts. After about 120 days they begin to deteriorate and lose their elasticity.

These worn out cells are consumed by large white cells which spit out the iron content, which in turn is picked up by transport molecules and taken to your bone marrow to be used in the manufacture of new red blood cells.

Every SECOND between two and three million new cells are released into the blood stream by the bone marrow.

No less important to our well-being are the white cells in our blood. When alien organisms invade our body, the immune system is alerted to the danger, and in the war against the invaders, the big guns are the bloods white cells. Like red blood cells they too are born in the bone marrow. Just like an army they form three distinct divisions : PHAGOCYTES and two types of LYMPHOCYTES called T cells and B cells.

During a disease millions of germs can be generated, and every one of these will have the same kind of antigen. Different diseases have different antigens. Before the T and B cells can attack these invaders they must have receptors that can bind to their particular antigens. Hence among the T cells and B cells, there has to be a vast array of receptors to cover each and every disease. Would it be up to the job and able to cope with the invasion ? The answer is furnished by Daniel E. Koshland Jnr., editor of the magazine SCIENCE. He says : "The immune system is designed to recognize foreign invaders. To do so it generates on the order of 10/11 (100,000,000,000) different kinds of immunological receptors so that no matter what the shape or form of the foreign invader there will be some complimentary receptor to recognize it and effect its elimination.

Amazing though it must sound, once the immune system has found a match for a disease and made an antibody, it can stay in the body for a very long time. This was recently discovered when examining blood from elderly survivors of the Spanish Flu epidemic of 1918. Scientists found "antibodies that still roam the body looking to strangle the old flu strain".

The INTERNATIONAL HERALD TRIBUNE went on to say : "Nine decades after history's most lethal flu faded away, survivors' bloodstreams still carry highly potent protection against the 1918 virus, demonstrating the remarkable durability of the human immune system".

The researchers that used these antibodies to make a vaccine were amazed at the immune system's memory.

It caused one of the researchers to say : "The Lord has blessed us with antibodies our whole lifetime. What doesn't kill you makes you stronger".

CHAPTER 19

For a moment lets now consider another vital organ, the LUNGS. Without them we couldn't survive as they extract life-giving oxygen from the air and distribute it throughout the whole body. The lungs are the two main organs located inside your ribcage. Your right lung has three sections or lobes, and your left lung has two. Each is somewhat independent from the other. At first glance the texture of the lung tissue appears to resemble sponge.

The most important muscle of respiration, the diaphragm, contributes to the constant inflation and deflation of the lungs. From the diaphragm your lungs extend all the way up into the base of your neck. A thin membrane covers each lung and also lines the inside of the chest wall. The space between the two membrane layers is filled with a lubricating fluid which enables the lungs and the rib cage to slide easily, without friction during respiration.

Various muscles, nerves, bones, cartilage, blood vessels, fluids, hormones and chemicals, all play a key role in the functioning of the lungs.

Before air reaches the lungs it has quite a journey to make. First the air flows from your nose or mouth into the pharynx or throat, which is used both for swallowing food and for breathing. To prevent food and drink entering your airways, the epigottis (a small movable lid) blocks the entrance when you swallow.

Now the air passes through the larynx, where your vocal chords are located. Next the air goes through the trachea or windpipe, then branches into two tubes known as the main bronchi. One bronchus enters the left lung, the other the right. Inside the lungs these tubes further divide into more branches. Because this branching keeps occurring again and again, the structure of the interior of the lungs resemble a tree with a trunk, branches and twigs. As the air keeps branching it gets thinner and thinner.

It then enters a network of miniature vessels called bronchioles, leading to even smaller ducts which send the air into some 300 million small air sacs known as alveoli. These are arranged in bunches and resemble tiny balloons. Its here that the systems airways end and the air reaches its final destination.

When it reaches this point, the air is contained within the extremely thin walls of alveoli. Each one is covered with a thin web of blood vessels known as pulminary capillaries. These are so narrow that only one red blood cell can pass through at a time and the walls are so thin that the carbon dioxide in the blood can seep through into the alveoli. The oxygen in turn passes in the opposite direction. It then exits the alveoli and is absorbed by the red blood cells. Each one of the cells travelling in single file, remains in the pulminary capillaries for about three-quarters of a second. This is plenty of time for the carbon dioxide and oxygen to change places. This movement is known as diffusion. The oxygenated blood then passes into larger veins in the lungs, eventually reaching the heart, from which its then pumped throughout the whole body. All told, it takes only about a minute for all the blood in your body to pass through this amazingly intricate system. Clever design don't you think?

Fortunately for us, the designer of the lungs made provision for quality control. Even the temperature of the air we breathe in is controlled. When the air is too cold, it is quickly heated to an adequate temperature. When too hot it is cooled down. When too dry, the walls of your nose, sinuses throat and other passageways are lined with a fluid called mucous. When you inhale dry air, moisture in the mucous evaporates into the air creating the required humidity which by the time it reaches the farthest point in your lungs, it has a relative humidity of 100%.

But the quality control doesn't end there, it also includes a sophisticated air filter which gets rid of infectious agents, toxic particles and other impurities.

The larger particles of dirt are collected in hairs and mucous membranes in our nose, which could be likened to our first line of defense against unwanted contaminates. Perhaps this is why we feel the need to clean out the unwanted debris on occasion.

Another way the lungs are protected is through the act of coughing and sneezing. When we cough, which is basically an abrupt expulsion of air, it rids the lungs of harmful substances when the lining of the respiratory tract becomes irritated. Coughing may also be a deliberate effort to clear the throat or bronchi.

Sneezing is an involuntary and violent rush of air through the mouth and nose. Nerve endings in the nose cause you to sneeze in order to get rid of irritating particles. In its way this too is a protection for the lungs.

Our breathing is caused by an automatic system with no conscious effort on our part. Even when we are asleep the lungs continue to work, extracting oxygen at a rate of 14 breaths per minute.

You can if you wish, temporarily override this automatic system by holding your breath. If this were not possible you would never be able to swim under water or escape a smoke filled room, during a fire. Of course you cannot bypass this system for very long (a few minutes at most) when your lungs will inevitably return to their automatic mode.

This automatic operation of inflation and deflation of the lungs is activated by special muscles. The control center is located in the brain stem. Here, receptors monitor the level of carbon dioxide in the body.

When it increases, messages are sent through a network of nerves which in turn activate the appropriate muscles of respiration.

Because of this our respiratory system is remarkably flexible. The lungs can keep up with even abrupt changes in your activity. For example during strenuous exercise, your body may use about 25 times as much oxygen and produce 25 times as much carbon dioxide as it does when its at rest. Almost instantaneously your lungs adjust the frequency and the depth of your breathing in order to match your constantly changing oxygen requirements.

Realistically, could such a sophisticated mechanism come about by chance ? You decide.

CHAPTER 20

What does the human body and the earth have in common ? To sustain life they both require a filter. The earth needs protection from the suns harmful rays which the ozone layer filters out.

The need for this protection was brought to my attention while on holiday in Tasmania, Australia. Because the hole in the ozone layer lies almost directly above the island, everyone there talks about the burning abilities of the sun over Tasmania. Now this may come as a surprise to those living on mainland Australia as they've been led to believe that the island doesn't enjoy an excess of hot days.

An example of the suns intensity on the island came up in conversation with a motel owner, whose close relatives had recently been on vacation there. I was told they came from Canberra and soon after they arrived went down to the beach. Within 20 minutes they were burnt, uncomfortably so ! Hence the need for an ozone layer without holes.

But what about the protection of the body, do we have a filter as well. Think about this : Many of the body's chemical processes release toxic substances and waste into your blood stream. If allowed to remain these would cause serious problems, even death. They must be continually filtered out and removed.

This is where your kidneys come in. As organs go they are fairly small, so how can they identify, isolate and remove harmful substances, yet at the same time ensure that vital elements needed to nourish the body remain.

The many elements making up your blood need to circulate throughout your entire body. As they do so they must repeatedly pass through your kidneys by means of large blood vessels called RENAL ARTERIES, one for each of your two kidneys. After entering the kidney this artery fans out into smaller vessels.

Finally the blood arrives at tiny clusters each having 40 tightly looped minute blood vessels. Each cluster, called a glomerulus is surrounded by a two layer membrane known as a BOWMAN'S CAPSULE. These two together make up the first part of your kidney's 'security gate' or NEPHRON which is the basic filtration unit of your kidney. You have over a million of these in your kidney and needless to say they are so small as to need a microscope to examine one.

Because blood cells and proteins in the bloodstream are indispensable, the first stage of filtration separates these from all the other elements . This task is carried out by the Bowmans Capsule. At this point filtration becomes more selective to be absolutely sure that nothing of value to your body escapes.

Connected to the Bowman's Capsule is a coiled tube called a CONVOLUTED TUBULE. The larger protein molecules and blood cells remain in the bloodstream and continue to flow freely. A watery mixture of dissolved useful molecules along with wastes, flows through the tubule. Along its inner wall are specialized cells which can recognize whats useful and whats not. The useful substances are efficiently plucked out by being reabsorbed into the tubule wall and eventually reentering the bloodstream. Now your blood has been filtered and cleansed it goes on to sustain life in your body.

What now happens to the unwanted fluid that remains in the tubule ?

Well, while it continues to flow along toward the larger collecting tubule, it receives from other cells in the wall secretions of ammonia, potassium, urea, uric acid and excess water. The final product we know as urine.

Conical structures known as RENAL PYRAMIDS deliver the urine to the renal pelvis, then leaves by way of the URETER, the tube connecting the kidney and the bladder. The bladder stores the urine until you are ready to expel it from your body.

That the two million nephrons in your kidney do a very impressive job was shown by the NEW ENCYCLOPAEDIA BRITANNICA. It states : "Nephrons......filter the entire five quart (5.5 liters) water content of the blood every 45 minutes".

At the end of this process, a normal healthy body can expel about 2.2 liters of waste in the form of urine every 24 hours. What a marvellous hardworking system !

Most doctors recommend we drink a little over two liters of water a day. According to former chairperson of the Urology Department of Long Island College Hospital New York, Dr. Godec : "Most people are dehydrated......you have to drink enough......most people don't".

Just drinking fluids is not enough. The sugar in fruit juices and sweetened drinks increases the body's need for water, and tea, coffee and alcohol actually cause the body to LOSE water. So if you want take care of your kidneys, be conscious of the need to drink just plain water.

In the last three chapters, I have dealt with the complexities of the brain, heart, lungs, kidneys and blood. There are of course many other body parts equally complex which time won't allow us to examine. But hopefully you get the point I'm trying to make : That where there is intricate design there must be a DESIGNER. There is nothing haphazard or accidental in the aforementioned organs.

At this point it would be appropriate to quote the words of a famous ancient king, that is King David of Israel. "He said : "I shall laud you because in a fear inspiring way I am wonderfully made. Your works are wonderful, as my soul is well aware". (Psalms 139 : 14)

How true. To deny the existence of an intelligent creator goes against all the facts and all reason and common sense. The universe, including our earth, the animals, birds, fishes, and man, all testify to a brilliant designer of genius proportions.

CHAPTER 21

Perhaps the biggest disagreements between the evolutionist and believers in creation, are over the age of man. How long have mankind been in existence ? That is the question.

Evolutionists claim over a million years —— Creation believers say 6000 years. Theres a big gap here !

To get to the bottom of this, we need to examine the different dating methods used and see whether they are fallible or infallible.

You are probably most familiar with the method known as radiocarbon dating. This method determines the amount of radioactive carbon (c14) left in bones, wood, charcoal, or some other once living object. C14 is an unstable element that decays. C14 is formed by the activity of cosmic rays on the earth's atmosphere. When man or animal eats the plants, the body absorbs the C14 from the plant. At death, this accumulation of C14 in the body stops, and what is already present continues to decay and is not replaced. In about 5,600 years the C14 is thought to be half gone, so it is said to have a half-life of that time.

Thus scientists take bone, wood, charcoal or other objects that were previously living and get an idea of their age by measuring the C14 left. If its half gone its considered to be about 5,600 years old. If three quarters gone its thought to be twice as old and so on. The method has its limitations because of its short half-life, so items over 50,000 years old cannot be dated by it.

When tested on old specimens connected with man what does the C14 reveal ? They show that the vast majority of such samples had radioactivity around the halfway point, well within 6,000 year span allowed for mans existence by the Bible. However, some objects that were dated seemed to indicate a longer period for mans existence than 6,000 years.

Does this mean these estimates prove the Bible wrong ?

For the answer its important to understand one thing about radiocarbon dating : It is based on assumptions. At a scientific conference held some years ago this fact was acknowledged.

SCIENCE magazine tells about it : "Throughout the conference emphasis was placed on the fact that laboratories do not measure ages, they measure sample activities. The connection between activity and age is made through a set of assumptions....... one of the main assumptions of C14 dating is that the atmospheric radiocarbon level has held steady over the age range to which the method applies". (51)

So what would happen if the C14 level in the atmosphere had not remained steady ?

SCIENCE DIGEST gives the answer : "It most certainly would ruin some of our carefully developed methods of dating things from the past...... if the level of carbon 14 was less in the past, due to a greater magnetic shielding from cosmic rays, then our estimates of the time that has elapsed since the life of the organism will be too long". (52)

To add weight to this comment SCIENCE YEAR said : "Scientists have found that the C14 concentration in the air and in the sea has not remained constant over the years, as originally supposed". (53)

An example of the infallibility of the C14 dating technique was exposed some years back when scientists tried to establish how long a pre-historic village had been occupied. The village was situated in Jarmo, northeastern Iraq. Eleven different determinations for its age covered a spread of 6,000 years. This estimate was totally at odds with all available archaeological evidence which showed it was not occupied for more than 500 consecutive years". (54)

SCIENCE magazine admitted the irresponsibility of C14 dating when it said : "Errors of shell radiocarbon dates may be as large as several thousand years". (55)

Ah you might say "But didn't I read a headline in the NEW YORK TIMES that declared : BONE FOUND IN KENYA INDICATES MAN IS 2.5 MILLION YEARS OLD ?" (56)

Because C14 dating is only good up to 50,000 years how did they date this bone from Africa ? They used what is called the potassium-argon method.

By measuring the content of potassium 40 and its decayed product argon 40, scientists try and determine the age of the rock, particularly volcanic rock. If the age of the rock above is determined then the bones underneath must be as old or older.

The problem is, this method is very uncertain in measuring the age of relatively recent volcanic rock. The reason : Radioactive potassium has a half-life of 1.3 billion years. In that vast time half of the potassium decays to become the gas argon. So measuring rocks just a few million years old is like trying to measure seconds on a clock that only has an hour hand.

NATURAL HISTORY magazine noted the extent of the problem when it said : "The potassium-argon dating method is increasingly inaccurate for dates of less than one million years. Consequently, there is a period during early and Middle Pleistocene times when dating human remains is difficult and uncertain". (57)

Potassium-argon dating of volcanic rock is made on a very weak assumption, that is, any volcanic activity dispelled all the argon originally in the molten lava. But if only a trace of argon remained, the clock would not be set back to zero and ages measured by it would be far to high.

A case in point was a find made in Africa by famous anthropologist Leakey. Note this observation in SCIENCE magazine : "The age of 1.75 million years......has been questioned on the basis of the possibility of the material being defective — for example, the material may have contained radiogenic argon at the time of crystallization or may have atmospheric contamination". (58)

Added to this, its been found that the potassium-argon ages do not always fall in the right sequence, in some cases the bed lying underneath gave an age younger than the bed lying above it.

The potassium in the earth has been generating argon all the time. When rock is melted in volcanic activity, every bit of argon would have to be boiled out for any reliable dating. Even if a minute trace remains it could cause errors amounting to millions of years. It would take only the tiniest trace of argon inherited from the melted rock to make a 5,000-year-old bed of volcanic rock look as though it were 1,750,000 or even 2,500,000 years old !

SCIENCE DIGEST showed just how unreliable this method can be when it said : "Through radioactive dating methods (potassium-argon) the age of the earth has been approximated at 4,500 million years. A new and higher figure — 6,500 million years — has now been given". (59)

That is a difference of 2 BILLION YEARS ! Why such a difference ? The same article explains that : "The new age for the earth may be the result of some overlooked factor in the potassium-argon technique".

From the foregoing discussion on dating methods it can be seen that none of them in any way disprove the 6,000 year age of mankind given by the Bible. Its true that animal fossils are much older but the Bible's account allows for this.

Many fossils have been buried under deep layers of earth and rock and cannot be accounted for purely due to volcanic activity. As a result, evolutionists, who, finding a deeply buried fossil where no evidence of volcanic activity existed, made the assumption it must be very old. Many of them believed that the earth's crust had not changed appreciatively since the first appearance of living things.

The facts show that earth's crust has not remained undisturbed. Giant upheavals have buried fossils far beneath terrestrial matter that was much older than the fossils it covered. Referring to this type of upheaval NEWSWEEK magazine makes the comment : "Catastrophism is a fighting word among geologists. It is a theory based on Divine intervention, and its adherents hold that the history of the earth and the life on it were moved by a series of disasters inspired by God — the last one, Noah's flood. It was the major line of thought for a few decades last century (19th) but a vigorous counterattack by the naturalists against the super-naturalists eventually pushed it aside".

"But now many geologists believe the counterattack may have been all to vigorous. In their haste to reject the hand of God, they passed over some solid evidence that could help improve their understanding of geology and evolution......There is evidence for example, that great expanses have been inundated within a matter of days. Such catastrophes were often followed by explosive development of different forms of life". (60)

The 1965 edition of SCIENCE YEAR took note of some drastic changes that have occurred in the earth's crust. It said : "The discovery of coal and fossil ferns in the transantarctic mountains....... was evidence of a warm climate in the past. Obviously there had been a reversal of climate". (61)

The above mentioned facts perfectly harmonize with the Bible's record of a global flood. Early in the Genesis account it mentions a great cloud layer that enveloped the whole earth. Because of this, the earth's climate was almost uniform. Even the North and South Pole were sub-tropical.

The Genesis account shows that God caused a "downpour upon the earth for 40 days and 40 nights". This vast quantity of water came from this massive cloud mass that enveloped the globe. Naturally, once most of this water had fallen to earth there would be a tremendous change in climate. It would now mean that the extreme northern and southern parts of the globe would be subject to freezing temperatures. Any living creatures in those areas would be caught up in an icy muck and snap frozen. Does this fit the facts ?

The SATURDAY EVENING POST had a special article dealing with this called : "Riddle of the frozen giants". It said : "About one seventh of the entire land surface of our earth, stretching in a great swathe round the Arctic Ocean, is permanently frozen...... the greater part of it is covered with a layer, varying in thickness from a few feet to a thousand feet, of stuff we call muck. This is composed of an assortment of substances, all bound together with frozen water, which becomes and acts as rock........

It is usually for the most part composed of fine sand and course silt, but it also includes a high proportion of earth and loam, and often also masses of bones or even whole animals in various stages of preservation or decomposition.........The list of animals that have been thawed out of this mess would cover several pages......The greatest riddle however, is when,why and how did all these assorted creatures, and in such absolutely countless numbers, get killed, mashed up and frozen into this horrific indecency.......?

These animal remains were not in deltas, swamps or estuaries, but were scattered all over the country.......But last and worst of all, many of these animals were perfectly fresh, whole and undamaged, and still either standing or at least kneeling upright...... Here is a really shocking — to our previous way of thinking — picture. Vast herds of enormous well fed beasts, not specifically designed for extreme cold, placidly feeding in sunny pastures, delicately plucking flowering buttercups at a temperature in which we would probably not even have needed a coat. Suddenly they were all killed without any visible sign of violence and before they could so much as swallow a last mouthful of food, and then were quick frozen so rapidly that every cell of their bodies is perfectly preserved". (62)

Thousands of these creatures have been found in Alaska, Canada and Siberia. Concerning the discoveries made in Alaska, French writer Francois Derrey makes this point in his book LA TERRE CETTE INCONNUE :

"The pits of Alaska are not unique. All over the world there have been charnel houses of this nature, piled high with the broken remains of thousands of animals".

An example of one of these "charnel houses" in an area far from the frozen wastes, was a discovery made in the 19th century of a huge array of bones and fossils in caves at Wellington N.S.W. Australia.

There were found the remains of giant marsupials, flightless birds, twice the size of today's emu's, diprotodons which resembled Wombats but were the size of cows, giant kangaroo's, marsupial lions, primitive koalas and many more. How they all ended up there in the caves remains a mystery. Many are broken, which strongly suggests they were washed into the cave system in a kind of muddy slurry. Its been suggested that a great flood may have carried them there.

So, science is discovering the facts, the truth of what the Bible shows that there were catastrophes that caused great climatic and terrestrial changes. One of these could well have been the Global deluge of over 4,000 years ago.

Returning to the main question of this chapter : How old is man ? There is one aspect of humans which make them unique. They keep written records. Its well to note why the last 6,000 years is called the 'historic' period of man. In this period it has been proven that man existed.

Science has the facts, records, documents, cities, monuments, writings and other artifacts to prove it. Before that period, man did not leave any evidence of his existence, hence that is why it is called "prehistoric".

The very idea of a "prehistoric" period for man is based solely on assumption, speculation. In a word it remains a THEORY with no facts in support.

That man is a relative newcomer, but equipped for rapid development, can be seen from such findings noted in the book NEW DISCOVERIES IN BABYLONIA ABOUT GENESIS. It says : "No more surprising fact has been discovered by recent excavation, than the suddenness with which civilization appeared in the world. This discovery is the very opposite to that anticipated. It was expected that the more ancient the period, the more primitive would excavators find it to be, until traces of civilization ceased altogether and aboriginal man appeared. Neither in Babylonia nor Egypt, the lands of the oldest known habitations of man, has this been the case".

This came as a blow to evolutionists who thought that ancient peoples would be found to have inferior intelligence to our modern selves. They hoped to find a gradual evolving of mans intelligence from mentally backward to highly intelligent. Alas, they were in for a rude awakening.

That the ancients already had a high brain capacity is shown by the following item in the NEW YORK TIMES about a discovery in Iraq : "Schoolboys of the little Sumerian country seat of Shadippur about 2000 B.C. Had a "textbook" with the solution of Euclid's classic triangle problem, seventeen centuries before Euclid.....It suggests that mathematics reached a stage of development about 2000 B.C. that archaeologists and historians of science had never imagined possible".

CONCLUSION

One reason why people believe evolution over creation is the misguided belief that the Bible says that God created the earth in six literal days. We all know that this is impossible, even for an almighty creator.

A careful examination reveals the "six days" are much longer than six days of 24 hours each. It may surprise some that the Bible does not disagree with scientists estimates of the age of the earth at billions of years.

The second verse of Genesis chapter one makes this statement : "Now the earth proved to be formless and waste and there was darkness upon the surface of the watery deep; and God's active force was moving to and fro over the surface of the waters". This description of the earth's condition was made BEFORE the first creative "day", hence allowing for any figure that might be put on the age of the earth.

Regarding the term 'day', the Hebrew word for this is 'yohm'. That it can have widespread meanings is shown by the book OLD TESTAMENT WORD STUDIES by William Wilson. To quote : "A day; it is frequently put for time in general, or for a long time; a whole period under consideration..

Day is also put for a particular season or time when any extraordinary event happens".

This last definition of a day would certainly fit the creative days, ("when an extraordinary event happens"). The Hebrew word day could mean summer or winter, harvest time, 24 hours, or 1000 years as in a day in Gods "timetable". It also allows for milleniums.

How often we too have used such expressions as "in my Father's day", or in "Shakespeare's day" or "Victoria's day". We know full well that we are not referring to a mere 24 hour day but a period of time covering a large portion of these individuals lives. In Queen Victoria's case we are thinking of her reign which covered a period of 64 years, even though we still call it Victoria's DAY.

I'm sure by this time you've concluded that this author is a firm believer in an intelligent Creator —— and you would be right. Perhaps by now you might think I've made a good case for creation. I sincerely hope so.

Whether or not you have changed any of your preconceptions since reading this book, it is my hope that in future, when some scientific boffin makes a statement like "Millions of years ago when life evolved in the oceans" you stop and ask : "How did he arrive at this conclusion. Wheres his proof" ?

This is the main purpose of this book....to get the public to QUESTION the assertions of evolutionists instead of accepting without query what they say. If their assertions are true, they will stand the glare of examination; if not, they will be shown to be based on an untruth in the absence of supportive facts. If their ideas are not rooted in fact they must be classed as unscientific.

As pointed out in the first chapter, I quoted the dictionary definition of the word MYTH. To refresh your memory it is : "A widely held but false belief or fictitious thing".

I made the claim that I would set out to prove that the evolution theory is worthy of this definition. Having read all my arguments so far, what do you think ? Is evolution a myth or a fact ?

You be the judge.

THE END

LIST OF REFERENCES

1. The Evolution of a Theory by C. Booker 1982 P.19.
2. Ibid P.199.
3. Scientific American August 1954 P.46.
4. The Selfish Gene by Richard Dawkins 1976 P. 16.
5. Ibid
6. Evolution from Space by F. Hoyle & C. Wickramasinghe P.27.
7. The Sciences "The Creationalists revival" by J.Gurin P.17.
8. Evolution : A Theory in Crisis by Michael Denton.
9. Evolution from Space P.8.
10. Ibid P. 20.
11. Life on Earth by David Attenborough P. 137.
12. The Reptiles by A. Carr P.36.
13. Ibid P.37.
14. Human Destiny by LeCompte Du Nouy P.72.
15.   "   "   "   "   "   "   "
16. Ibid P. 295.
17. Lucy :   The beginnings of Humankind by D.Johanson & M.Edey P.27.
18. Science Digest "The Water People" by L. Watson P.44.
19. Man, God and Magic by I. Lissner P. 304.
20. World Book Encyclopaedia (Under heading Neanderthal Man).
21. Scientific American Dec. 1966 P.32.
22. The Origin of the Species by Charles Darwin 1902 edition P.54.
23. Ibid pp. 83, 88, 91, 92.
24. New Scientist January 15 1981 P.129.

25. The Synthetic Origin of Species by Heribert Nilson 1953 P.1212.
26. Red Giants and White Dwarfs by Robert Jastrow 1979 P.97.
27. Evolution from Space P8.
28. Natural History "Darwin and the Fossil Record" by A.S.Romer P466/7.
29. New York Times Oct. 3 1982 section 1 P.49.
30. DISCOVER "The tortoise or the Hare" ? By J. Gorman Oct 1980 P.89.
31. Should Evolution be taught ? By John N. Moore 1970 pp. 9, 14.
32. Liberty September/October 1975 P.14.
33. The Enchanted Loom—Mind in the universe.
34. The New England journal of Medicine September 13 1973 V289,P577.
35. The Center of Life by L.L. Larison Cudmore 1977 pp 23,24.
36. How Life Learned to Live by Helmut Tributsch 1982 P.204.
37. The Atlantic Monthly "Debating the Unknowable" by L.Thomas 7/81.
38. How Life learned to Live P.68.
39. The Origin of the Species by C.Darwin Mentor ed. 1958 P228.
40. The Great Evolution Mystery by G.Rattray Taylor 1983 pp. 221/2.
41. "   "   "   "   P.221.
42. Neuronal Man by Dr. Jean-Pierre Changeux.
43. The Brain : The Last Frontier by Richard M. Restak.
44. Ibid "The development of the brain" by W.Maxwell Cowan P.131.
45. Ibid "The Brain" by David H. Hubel P.52.
46. The Brain : The Last Frontier P.158.
47. Cosmos by Carl Sagan 1980 P.278.
48. Scientific American "Specializations of the Human Brain". Sept 1979.
49. Ibid pp.227-229.
50. Readers Digest "Thoughts of a Brain Surgeon" by R.J.White Sept. '78.
51. Science magazine Dec. 10 1965 P.1490.
52. Science Digest Dec. 1960 P.19.
53. Science Year 1966 P.193.

54. Science magazine Dec 11, 1959 P.1630.
55. Science   "   Aug.16, 1963 (Ibid P.634).
56. New York Times Jan 14 1967 P.1.
57. Natural History February 1967 P.58.
58. Science magazine April 2 1965 P.73.
59. Science Digest Dec 1962 P.35.
60. Newsweek Dec. 23 1963 P.48.
61. Science Year 1965 pp.215, 217.
62. Saturday Evening Post January 16 1960 pp.39, 82, 83.